ONI 226
CONFIDENTIAL

ALLIED LANDING CRAFT AND SHIPS

World War II Recognition Guide

Publications and Distribution Branch—Division of Naval Intelligence

by **OFFICE OF NAVAL INTELLIGENCE**

ISBN #978-1-940453-16-3

ISBN #978-1-940453-16-3

INTRODUCTION

ONI 226—Allied Landing Craft and Ships—is designed to facilitate recognition of and to make available pertinent data regarding the characteristics of United States and British vessels which may engage in amphibious operations.

★

Information of a nontechnical nature necessary for the recognition and use of landing craft, ships, and vehicles is provided in condensed form. Figures given are based on the best obtainable information. Experience has shown, however, that speed, endurance, and similar characteristics of these vessels may vary under operational conditions, and data given here should not be accepted when in conflict with experience in the field.

★

The following general principles form the basis for names and designated letters used in connection with equipment described herein.

(a) The term Landing Craft is generally applied to non-oceangoing vessels of less than 150 feet length over-all, designed for use in landing operations. The designation LC (Landing Craft) is therefore used, with appropriate modification to indicate particular types.

(b) At the date of issue of this publication, oceangoing vessels of over 200 feet length over-all, specifically designed for landing operations, fall within the category of Landing Ships. LS is used as a prefix in special designations of these types.

(c) Small units used in landing operations which are capable of use on land or water are known as Landing Vehicles. The letters LV are used as a prefix for all designations referring to these types.

★

It will be noted that a number of type names, on which no data is given, are listed within the various categories. These type names have been officially established for vessels which are in either the design or the construction stage. As further information becomes available, it will be forwarded to holders of ONI 226.

Suggestions for improvement or changes should be transmitted to the Division of Naval Intelligence.

NAVY DEPARTMENT
Office of the Chief of Naval Operations
Washington
7 April 1944

1. ONI 226—Allied Landing Craft and Ships—supersedes FTP-207—Characteristics of British and United States Landing Craft, Landing Ships, and Landing Vehicles. It is issued for the use of United States and British units and personnel engaged in amphibious operations.

2. Holders of ONI 226 will safeguard the security of this publication in accordance with the provisions of Article 76, U. S. Navy Regulations. When no longer required, this publication will be destroyed by burning. No report of destruction is required.

R. E. Schuirmann
Rear Admiral, U. S. Navy
Director of Naval Intelligence

Op-16-P-2 A10-1 ONI 226 Serial No. 0547116 CONFIDENTIAL

LIST OF DESIGNATIONS

Publication and Distribution Branch—Division of Naval Intelligence

Abbreviated Designation	Full Designation	Former Designation
CRAFT		
LCA	Landing Craft, Assault	ALC
LCA(HR)	Landing Craft, Assault (Hedgerow)	
LCC	Landing Craft, Control	
LCE	Landing Craft, Emergency Repair	
LCF	Landing Craft, Flak	BPC
LCG	Landing Craft, Gun	
LCH	Landing Craft, Headquarters	
LCI	Landing Craft, Infantry	Giant Raiding Craft, APY
LCM	Landing Craft, Mechanized	MLC, TL, WL
LCN	Landing Craft, Navigational	
LCP	Landing Craft, Personnel	28′ RC, CRC, R Craft, Y, T
LCP(N)	Landing Craft, Personnel (Nested)	LCP(P)
LCP(R)	Landing Craft, Personnel (Ramped)	
LCP(SY)	Landing Craft, Personnel (Survey)	
LCP(U)	Landing Craft, Personnel (Utility)	
LCR	Landing Craft, Rubber	
LCS	Landing Craft, Support	LSC, HSC
LCS(R)	Landing Craft, Support (Rocket)	
LCT	Landing Craft, Tank	TLC, MTL
LCT(A)	Landing Craft, Tank (Armored)	
LCT(R)	Landing Craft, Tank (Rocket)	
LCV	Landing Craft, Vehicle	VLC, YR, TR
LCVP	Landing Craft, Vehicle Personnel	
LCW	Landing Craft, Air Propelled	

Abbreviated Designation	Full Designation	Former Designation
SHIPS		
LSB	Landing Ship, Bombardment	
LSC	Landing Ship, Carrier (Derrick-Hoisting)	
LSD	Landing Ship, Dock	
LSE	Landing Ship, Emergency Repair	
LSF	Landing Ship, Fighter Direction	
LSG	Landing Ship, Gantry	
LSH	Landing Ship, Headquarters	
LSI	Landing Ship, Infantry	
LSI(H)	Landing Ship, Infantry (Hand-Hoisting)	
LSM	Landing Ship, Medium	
LSP	Landing Ship, Personnel	
LSS	Landing Ship, Stern-chute	
LST	Landing Ship, Tank	
LSV	Landing Ship, Vehicle	
BARGES		
LBE	Landing Barge, Emergency Repair	
LBF	Landing Barge, Flak	LBG, LBS
LBK	Landing Barge, Kitchen	
LBO	Landing Barge, Oiler	
LBW	Landing Barge, Water	
LBV	Landing Barge, Vehicle	AA, BB, CC
VEHICLES		
LVT	Landing Vehicle, Tracked	
LVT(A)	Landing Vehicle, Tracked (Armored)	

LANDING CRAFT TYPES

*Major Operational Types

PERSONNEL LANDING CRAFT

		Builder
*LCA	Landing Craft, Assault	UK
LCI(S)	Landing Craft, Infantry (Small)	UK
*LCI(L)	(Large)	US
LCP(1)	Landing Craft, Personnel (Mark 1)	UK
LCP(2)	(Mark 2)	UK
LCP(S)	(Small)	UK
LCP(M)	(Medium)	UK
*LCP(L)	(Large)	US
LCP(N)	(Nested)	US
*LCP(R)	(Ramped)	US
LCP(U)	(Utility)	UK
*LCR(S)	Landing Craft, Rubber (Small)	US
*LCR(L)	(Large)	US
LCW	Landing Craft, Air Propelled	UK

VEHICLE AND TANK LANDING CRAFT

		Builder
LBV(1)	Landing Barge, Vehicle (Mark 1)	UK
LBV(2)	(Mark 2)	UK
LCM(1)	Landing Craft, Mechanized (Mark 1)	UK
LCM(2)	(Mark 2)	US
*LCM(3)	(Mark 3)	US
LCM(4)	(Mark 4)	UK
LCM(5)	(Mark 5)	UK
LCM(6)	(Mark 6)	US
LCM(7)	(Mark 7)	UK
LCT(1)	Landing Craft, Tank (Mark 1)	UK
LCT(2)	(Mark 2)	UK
LCT(3)	(Mark 3)	UK
LCT(4)	(Mark 4)	UK
*LCT(5)	(Mark 5)	US
*LCT(6)	(Mark 6)	US
LCT(8)	(Mark 8)	UK
LCT(9)	(Mark 9)	UK
*LCV	Landing Craft, Vehicle	US
LCV(2)	(Mark 2)	UK
*LCVP	Landing Craft, Vehicle, Personnel	US
Z-Lighter		UK

SUPPORT AND COMMAND CRAFT

		Builder
LBF	Landing Barge, Flak	UK
LCA(HR)	Landing Craft, Assault (Hedgerow)	UK
*LCC(1)	Landing Craft, Control—(Mark 1)	US
*LCC(2)	(Mark 2)	US
LCC		
LCF(2)	Landing Craft, Flak—(Mark 2)	UK
LCF(3)	(Mark 3)	UK
LCF(4)	(Mark 4)	UK
LCG(M)	Landing Craft, Gun (Medium)	UK
LCG(L)(3)	(Large)—(Mark 3)	UK
LCG(L)(4)	(Large)—(Mark 4)	UK
LCH	Landing Craft, Headquarters	UK
LCI	Landing Craft, Infantry (Gunboat)	US
LCN	Landing Craft, Navigational	UK
LCP(SY)	Landing Craft, Personnel (Survey)	UK
*LCS(S)(1)	Landing Craft, Support (Small)—(Mark 1)	US
LCS(S)(2)	(Small)—(Mark 2)	US
LCS(M)(1)	(Medium)—(Mark 1)	UK
LCS(M)(2)	(Medium)—(Mark 2)	UK
LCS(M)(3)	(Medium)—(Mark 3)	UK
LCS(L)(1)	(Large)—(Mark 1)	UK
LCS(L)(2)	(Large)—(Mark 2)	UK
*LCS(L)(3)	(Large)—(Mark 3)	US
LCS(R)	(Rocket)	UK
LCT(A)	Landing Craft, Tank (Armored)	UK
LCT(R)	Landing Craft, Tank (Rocket)	UK
PGM	Motor Gunboat	US

AMPHIBIOUS VEHICLES

		Builder
LVT(1)	Landing Vehicle, Tracked—(Mark 1)	US
*LVT(2)	(Mark 2)	US
*LVT(3)	(Mark 3)	US
*LVT(4)	(Mark 4)	US
*LVT(A)(1)	(Armored)—(Mark 1)	US
*LVT(A)(2)	(Armored)—(Mark 2)	US
LVT(A)(3)	(Armored)—(Mark 3)	US
*LVT(A)(4)	(Armored)—(Mark 4)	US

AMPHIBIOUS VEHICLES—Con.

		Builder
*DUKW	2½-Ton, 6 x 6, Amphibian Truck	US
*JEEP	¼-Ton, 4 x 4, Amphibian Truck	US
TERRAPIN 1		UK
TERRAPIN 2		UK

REPAIR AND SUPPLY CRAFT

		Builder
LBE	Landing Barge, Emergency Repair	UK
LBK	Landing Barge, Kitchen	UK
LBO	Landing Barge, Oiler	UK
LBW	Landing Barge, Water	UK
*LCE	Landing Craft, Emergency Repair	UK

SHIPS

		Builder
LSB	Landing Ship, Bombardment	UK
LSC	Landing Ship, Carrier (Derrick-Hoisting)	UK
*LSD	Landing Ship, Dock	US
LSE	Landing Ship, Emergency Repair	UK
LSF	Landing Ship, Fighter Direction	UK
LSG	Landing Ship, Gantry	UK
LSH	Landing Ship, Headquarters	UK
*LSI(H)	Landing Ship, Infantry (Hand-Hoisting)	UK
LSI(S)	(Small)	UK
LSI(M)	(Medium)	UK
LSI(L)	(Large)	UK
*LSM	Landing Ship, Medium	UK
LSP	Landing Ship, Personnel	UK
LSS	Landing Ship, (Stern-chute)	UK
LST	Landing Ship, Tank	UK
LST(1)	(Class 1)	UK
*LST		US
LST(3)	(Class 3)	UK
LST(4)	(Class 4)	UK
LSV	Landing Ship, Vehicle	US
AGC	Amphibious Force Flagship	US
*APc	Coastal Transport (Small)	US
*APD	High Speed Transport (Destroyer)	US
*ARL	Landing Craft, Repair Ship	US

U. S. AND BRITISH LANDING CRAFT

Type	Capacity	Length	Beam	Draft	Weight or displ.	Speed (Knots)	Endurance (Miles)	Armament	Armor	Crew
BARGES										
LBE, Landing Barge, Emergency Repair	3 ton workshop lorry	75′–85′	18′–23′	4′0″–4′6″ mean (loaded)		6 max.	360 @ 5 kts.	One 20 mm. Oerlikon	10-lb. steel plating over cabin	18–24
LBF, Landing Barge, Flak	Crew only and 16 Army	60′–90′	19′–24′	5′ mean		6–7 max.	300 @ 5 kts. max. (cont.)	Two 40 mm. Bofors, two 20 mm. Hispano or 2 twin Lewis guns.	3″ concrete bottom and sides, 4′6″ up	6
LBK, Landing Barge, Kitchen	Provisions for 900 men for 1 week	79′	21′	3′6″ est.			300 @ 5 kts. est.	None	2½″ plastic to fuel tanks, engines, and steering shelter	25
LBO, Landing Barge, Oiler	40 tons in tank	80′–90′	20′–23′	4′6″–6′0″ mean (loaded)		6 max.	360 @ 5 kts.	1 twin 303 Lewis gun	2½″ plastic on fuel, gas tanks, and steering shelter; 5-lb. plating over engine; 3″ concrete sides to hold	
LBW, Landing Barge, Water	40 tons in tank	80′–90′	20′–23′	4′6″–6′0″ mean (loaded)		6 max.	360 @ 5 kts.	1 twin .303 Lewis gun	2½″ plastic on fuel gas tanks, and steering shelter; 5-lb. plating over engine; 3″ concrete sides to hold	
LBV(1), (2), Landing Barge, Vehicle (Small), (Mark 1, 2)	100 tons, (approx.); two 3-ton or three 15-cwt. lorries	70′	18′	4′ to 4′6″ mean (loaded)		3–5 max. (loaded)	360 @ 5 kts. (Mk. 2)		Concrete hull sides; 2½″ plastic to gas tanks when fitted; 2 portable steel bulkheads inboard of ramp	

Type	Capacity	Length	Beam	Draft	Weight or Displ.	Speed (Knots)	Endurance (Miles)	Armament	Armor	Crew
BARGES—Con.										
LBV(1), (2), Landing Barge Vehicle (Medium)—(Mark 1, 2)	150 tons, (approx.); two 6-pdr. guns and 3 lorries; or two 3-ton lorries, one 15-cwt. lorry, and two 30-cwt. lorries	78′	20′	4′ to 4′6″ mean (loaded)		3–5 max. (loaded)	360 @ 5 kts. (Mk. 2)		Concrete hull sides; 2½″ plastic to gas tanks when fitted; 2 portable steel bulkheads inboard of ramp	
LBV(1), (2) Landing Barge, Vehicle (Large)—Mark (1),(2)	200 tons (approx.); two 6-pdr. guns and 3 lorries; or two 3-ton lorries, one 15-cwt. lorry, and two 30-cwt. lorries	82′	23′	4′ to 4′6″ mean (loaded)		6 max. (loaded)	360 @ 5 kts. (Mk. 2)		Concrete hull sides; 2½″ plastic to gas tanks when fitted; 2 portable steel bulkheads inboard of ramp	
Z-Lighter	Six 15 ton tanks or ten 6-wheeled lorries or nine 4-wheeled lorries and 95 tons fresh water	134′ o. a.	30′	2′2″ for'd, 4′ aft, 3′ (mean) with 80 ton load			236 miles			
CRAFT										
LCA, Landing Craft, Assault	35 troops and 800 lb. equipment	41′6″	10′	1′9″ for'd 2′3″ aft (loaded)	13 tons (max.)	6 (loaded); 10 max. (light)	50–80 @ 7 kts. (loaded) 65 @ 10 kts.	1 Bren gun; two .303 Lewis guns; two 2″ mortars	10-lb. DIHT to bulkheads, hull sides; 7.8-lb. DIHT to decks	4 plus 1 off. per 3 craft
LCA(HR) Landing Craft, Assault (Hedgerow)	Crew only	41′6″	10′	1′9″ for'd 2′3″ aft (loaded)	13 tons (max.)	10 max.	50–80 @ 7 kts. (loaded) 65 @ 10 kts.	24 mortars	10-lb. DIHT to bulkheads, hull sides; 7.8-lb. DIHT deck	4 plus 1 off. per 3 craft.
LCC, Landing Craft, Control		112′	18′6″	5′ (max.)	75.5 tons	20 max. 17 kts. cruising		One 3-pdr., 20 mm., 1 Y gun, 4 M. G.		

Type	Capacity	Length	Beam	Draft	Weight or Displ.	Speed (Knots)	Endurance (Miles)	Armament	Armor	Crew
CRAFT—Con.										
LCC(1), Landing Craft, Control—(Mark 1)	Crew only	56′	13′7″	3′11½″ max.	30 tons	13.5 max.	500 @ 10 kts.; 240 @ max.	3 twin .50 cal. M.G.	¼″ STS on bridge and cockpits	14
LCC(2), Landing Craft, Control—(Mark 2)	Crew only	56′	15′	3′11″ max.	30 tons	13.5	500 @ 10 kts.; 250 @ max.	2 twin .50 cal. M.G.	¼″ STS on bridge and cockpits	14
LCE, Landing Craft, Emergency Repair	Crew and repair gear	36′3″	10′10″	3′	23,500 lb. (max.)	9	80 @ max. speed	None	10-lb. DIHT on engine sides, bulkheads, hull sides	10
LCF(2), Landing Craft, Flak—(Mark 2)	Crew only; 68 gun crew	159′11″	31′	3′6″ for'd 7′6″ aft (loaded)	455 tons	11.5 (loaded) max.	2,000 @ 10 kts.	2 twin 4″ DP, three 20 mm.	20-lb. DIHT or 2½″ plastic to bridge, gun deck	13
LCF(3), Landing Craft, Flak—(Mark 3)	Crew only; 50 gun crew	195′9″	31′	2′5″ for'd 6′9″aft (light)	440 tons (light)	9½ max.	2,700 @ 9½ kts.	Eight 2-pdr. pompoms and four 20 mm. (LCF 3–6), or four 2 pdr. pompoms and eight 20 mm. (LCF 7–18)	15-lb. DIHT to gun shields and positions; plastic on bridge	12
LCF(4), Landing Craft, Flak—(Mark 4)	Crew only; 50–54 gun crew	187′3″	38′8″		370 tons (light)	9.5 max. (loaded)	1,100 @ 8 kts.	Four 2 pdrs. Eight 20 mm.	15-lb. DIHT to gun-shields and positions; plastic on bridge	12
LCG(L)(3), Landing Craft, Gun (Large)—(Mark 3)	Crew only; 23 gun crew	192′	31′	3′6″ for'd 6′0″ aft	500 tons (loaded)	10½ max.	2,700 @ 9 kts.	Two 4″.7 QF Naval guns; two 20 mm. or two 2 pdr. pompoms	25-lb. DIHT to side of gun deck, blast screens, sides of magazines	15

Type	Capacity	Length	Beam	Draft	Weight or Displ.	Speed (Knots)	Endurance (Miles)	Armament	Armor	Crew
CRAFT—Con.										
LCG(L) (4), Landing Craft, Gun (Large)—(Mark 4)	Crew only; 23 gun crew	187′3″	38′8″	3′6″ for'd 6′0″ aft	500 tons (loaded)	10	1,100 @ 8 kts.	Two 4″.7 QF Naval guns; two 20 mm. or two 2 pdr. pompoms	25-lb. DIHT to side of gun deck, blast screens, sides of magazines	15
LCG(M), Landing Craft, Gun (Medium)	Crew only	154′6″	22′4″	6′3½″(mean) (loaded) 5′6″ operating	270 tons (light)	13½		Two 17 pdr. or 25 pdr. Army guns; two 20 mm; 2 smaller M.G.	2″ NC hull sides, control station; 1″ NC deck over engines and mag.; 15-lb. DIHT to gun pos.-tions, conning tower	30
LCH, Landing Craft, Headquarters	(Statistics same as LCI (L))									
LCI (L)—(1-350), Landing Craft, Infantry (Large)	6 officers, 182 enlisted men or 75 tons cargo	158′5½″	23′3″	2′8″ for'd 4′10″ aft (landing) 3′1½″(mean)	380 tons (loaded) 216 tons (light) 234 tons (landing)	16 (max. loaded)	4,000 @ 12 kts. (beaching draft); 500 @ 15 kts. or 1,500 @ 12 kts.	Four 20 mm.	2″ plastic on guns, conning and pilot-house sides	24
LCI(L)—(351—), Landing Craft, Infantry (Large)	9 officers, 196 enlisted men, 32 tons cargo	158′5½″	23′3″	3′ for'd, 5′0″ aft (landing) 5′8″ for'd, 5′8″ aft (loaded)	250 tons (landing) 387 tons (loaded)		8,000 @ 12 kts.	Five 20 mm.	2″ plastic on guns, conning and pilot-house sides	28

Type	Capacity	Length	Beam	Draft	Weight or Displ.	Speed (Knots)	Endurance (Miles)	Armament	Armor	Crew
CRAFT—Con.										
LCI(S), Landing Craft, Infantry (Small)	102 fully equipped troops and 18 bicycles	105'1"	21'4"	2'10" for'd 3'8" aft (loaded)	63 tons (light) 100 tons (loaded)	15 max. 13¼ max. (cont.)	550 @ 15 kts.; 700 @ 12½ kts.	Four 20 mm. (2 fitted); two .303 Lewis guns	10-lb. DIHT to deck, sides, gun positions, generator, and for'd bulkhead; ¼" plating to bridge	17
LCM (1), Landing Craft, Mechanized—(Mark 1)	Tanks or trucks of less than 16 tons total, or 100 men or mixed loads; some units can carry 17¼ tons increasing mean draft 1 inch	44'8"	14'0"	1'6" for'd 4'0" aft (loaded), 1'4" for'd, 2'9" aft (light)	35 tons (loaded) 21 tons (light)	7.5	56 @ 7.5 kts.	Two .303 Lewis guns	8 or 10-lb. bulwarks, deck, control, and fuel tanks	6 plus 1 off. per 3 craft
LCM (2), Landing Craft, Mechanized—(Mark 2)	1 light tank (13½ tons) or 30,000 lb. cargo, or 100 troops	45'	14'1"	3' (mean)	58,000 lb.	7.5	75 @ 7.5 kts. max.	Two .50 cal. M.G.	¼" on control station	4
LCM (3), Landing Craft, Mechanized—(Mark 3)	One 30-ton tank or 60,000 lb. cargo or 60 troops; Bureau type carries 120,000 lb. cargo	50'	14'1"	3' for'd, 4' aft (light)	52,000 lb. (light) 52 tons (loaded)	11 max. 8 (loaded)	850 @ 6¼; 140 @ 11 kts.	Two .50 cal. M.G.	¼" HTS to control station; British fitting bulletproof mattresses	4
LCM (6), Landing Craft, Mechanized—(Mark 6)	One 30-ton tank or 60,000 lb. cargo or 60 troops; Bureau type carries 120,000 lb. cargo	56'	14'1"	3' for'd, 4' aft (light)	52,000 lb. (light)	11	850 @ 6¼; 150 @ 11 kts.	Two .50 cal. M.G.	¼" HTS to control station; British fitting bulletproof mattresses	4 plus 1 off. per 3 craft (6 British)
LCM (7), Landing Craft, Mechanized—(Mark 7)	One 30-ton tank Hold—35' x 13'6" Ramp—13'6" wide			3'6" (max.)	23 tons (light)	11	200 @ 11 kts.			

Type	Capacity	Length	Beam	Draft	Weight or Displ.	Speed (Knots)	Endurance (Miles)	Armament	Armor	Crew
CRAFT—Con.										
LCN, Landing Craft, Navigational	30–36 troops or 6,700-8,100 lb. cargo; fuel space may reduce this figure	36′	10′9″	2′6″ aft (light) 3′6″ aft (loaded)	18,000 lb. (loaded)	11 max. (loaded)	80—early types; 120 @ 11 kts. 200 on later types	Two .30 cal. M.G.; early type unarmed or one .303 Lewis gun	Three 1/4″ transverse bulkheads	3 plus 1 off. per 3 craft
LCP(1), Landing Craft, Personnel—(Mark 1)	18 men	20′	6′		2,000 lb. (light)		Towed			2
LCP(2), Landing Craft, Personnel—(Mark 2)	18 men	25′6″	6′3″		4,000 lb. (light)	10	30 @ 10 kts. max. (loaded)			3
LCP(L), Landing Craft, Personnel (Large) (U. S. Type)	30–36 troops or 6,700-8,100 lb. cargo; fuel space may reduce this figure	36′8″	10′10″	2′6″ aft (light) 3′6″ aft (loaded)	18,000 lb. (loaded); 13,000 lb. (light)	8 max. (loaded)	50 (gasoline) 130 (Diesel)	Two .30 cal. M.G.	Three transverse 10-lb. bulkheads	3
(U. K. Type)	25 equipped troops or 6700-8100 lb. cargo	36′8″ (o. a.)	10′10″	2′6″ aft (light) 3′6″ aft (loaded)	18,000 lb. (loaded); 13,000 lb. (light)	9–11 max. (loaded)	120 @ full speed (loaded)	One .303 Lewis gun	Some being fitted with 10-lb. plating	3 men, 1 off. per 3 craft
LCP(M), Landing Craft, Personnel (Medium)	20 troops	38′11″	10′1″	2′1″ for'd 1′7″ aft (loaded)	4½ T. (light) 7¼ T. (loaded)	7½ max. (loaded)	112 @ 7½ kts.	One .303 Lewis gun	None	3 plus 1 off. per 3 craft
LCP(S) Landing Craft, Personnel (Small)	20 troops, or 6 litters and 10 troops, or 25 unequipped	29′6″	8′9″	1′4″ mean (light) 1′9″ mean (loaded)	3 tons (light) 5½ tons (loaded)	8 max. (loaded)	120 @ 8 kts.	One .303 Lewis gun	None	3

U. S. AND BRITISH LANDING CRAFT

Type	Capacity	Length	Beam	Draft	Weight or Displ.	Speed (Knots)	Endurance (Miles)	Armament	Armor	Crew
CRAFT—Con.										
LCP(N) Landing Craft, Personnel (Nested)	13 troops	32′	7′10″	10″ (loaded) designed	1,400 lb. w/o engine	10 – 12 m.p.h.		None	None	2
LCP(R) Landing Craft, Personnel (Ramped)	30–36 troops or 6,700–8,100 lb. cargo; fuel space may reduce capacity	35′10″	10′9″	2′6″ aft (light) 3′6″ aft (loaded)	13,000	11 max. (loaded)	80 or 200	Two .30 cal. M.G.; early U.K. type unarmed or one .303 Lewis gun	Three 1/4″ plate; transverse bulkheads	3
LCP(SY), Landing Craft, Personnel (Survey)	(No data available)									
LCP(U), Landing Craft, Personnel (Utility)	30 troops (fully equipped)			3′6″ (max.) aft	10 tons (light)	15 kts. designed; 12 min. designed	150 @ max. speed (loaded)			4
LCR(L), Landing Craft, Rubber (Large)	10 troops	16′	8′		395 lb. (light) 474 lb. with motor	3 1/2–4 1/2		One .30 M.G.		None
LCR(S), Landing Craft, Rubber (Small)	7 troops	12′5″	5′11″		210 lb. (light) 277.4 with motor	3 1/2–4 1/2 with engine				None
LCS(L) (1), Landing Craft, Support (Large)—(Mark 1)	Crew only	46′11″	12′7″	2′4″ for'd 4′3″ aft (loaded) 2′2″ for'd 3′11″ aft (light)	24.5 tons (loaded) 20 tons (light)	10 max.	100 @ 10 kts.	One 2 pdr. turret; 1 twin .50 M.G.; one 4″ smoke mortar; two .303 Lewis guns	10-lb. DIHT to deck top of turret; 12-lb. to top and 20 lbs. to sides of steering position; 15-lb. to sides of hull	13

Type	Capacity	Length	Beam	Draft	Weight or Displ.	Speed (Knots)	Endurance (Miles)	Armament	Armor	Crew
CRAFT—Con.										
LCS(L) (2), Landing Craft, Support (Large)—(Mark 2)	Crew only	105'1"	21'5"	3'8" for'd 3'4" aft (loaded) 2'8" for'd 3'0" aft (light)	116 tons (loaded) 84 tons (light)	14¾	700 @ 12½ kts.	One 6 pdr. turret; two 20 mm.; 1 twin .50 M.G.; one 4" mortar and two .303 Lewis	10-lb. DIHT to deck, hull sides, and gun positions; ¼" plating on bridge	25
LCS(L)(3), Landing Craft, Support (Large)—(Mark 3)	Crew only	158'5"	23'3"	4'6" for'd 5'10" aft	227 tons (light) 383 tons (loaded) est.			One 3"/50 DP; 2 twin 40 mm.; four 20 mm.	10-lb. splinter shielding to gun positions.	73
LCS(M) (1), Landing Craft, Support (Medium)—(Mark 1)	Crew only	41'2"	10'0"	1'8" for'd 1'11" aft (loaded) 1'5" for'd 1'9" aft (light)	10.8 tons (loaded) 8.9 (light)	10 max.	60 @ 10 kts.	Two .50 M.G.; two .303 Lewis; one 4" smoke mortar; or one 20 mm.	10-lb. plating to sides, parts of deck, and bulkheads; 20-lb plating to conning tower	11
LCS(M) (2), Landing Craft, Support (Medium)—(Mark 2)	Crew only	41'2"	10'0"	1'6" for'd 2'5" aft (loaded) 1'4" for'd 2'2" aft (light)	12.5 tons (loaded) 10.5 tons (light)	9 max.	60 @ 9 kts.	Two .50 M.G.; two .303 Lewis; one 4" smoke mortar; or one 20 mm.	10-lb DIHT to sides, parts of deck, and bulkheads; 20-lb plating to conning tower	11
LCS(M)(3), Landing Craft, Support (Medium)—(Mark 3)	Crew	41'2"	10'1"	1'9" for'd 2'3" aft (loaded) 1'6" for'd, 2'0" aft (light)	12.8 tons (loaded) 11.4 tons (light)	9 max.	90 @ 9 kts.	Two .50 twin M.G. in turret; two .303 Lewis; one 4" smoke mortar	10-lb. DIHT on hull sides, deck, bulkheads, conning tower, and ammunition hoist	10
LCS(R), Landing Craft, Support (Rocket)	(No data avialable)									

Type	Capacity	Length	Beam	Draft	Weight or Displ.	Speed (Knots)	Endurance (Miles)	Armament	Armor	Crew
CRAFT—Con.										
LCS(S)(1), Landing Craft, Support (Small) —(Mark 1)	3–4 plus crew and gunners	36'8''	10'10''	3'6'' (light)	20,000 lbs. (light) 22,000 lbs. standard	12	115 @ full speed	Two .50 cal. or three .30 cal. or one .50 cal. and two .30 cal. and 2 rocket projectors	¼'' plating to bridge	6
LCS(S)(2), Landing Craft, Support (Small) —(Mark 2)	3–4 plus crew and gunners	36'8''	10'10''	3'6'' (light)	23,000 lbs.	10	135 @ 11½ kts. max.	One .50 cal. twin mounts; two .30 cal. M. G.; two 12-rocket projectors	¼'' plating to bridge	6
LCT (1), Landing Craft, Tank—(Mark 1)	Three 40-ton tanks, or six 25-ton tanks, or six 16-ton tanks, or 250 tons cargo	152'	29'	3' for'd, 5'9'' aft (loaded)	226 tons (light)	10 max.	900 @ 10 kts. max.	Two 2-pdrs.		12
LCT (2), Landing Craft Tank—(Mark 2)	Three 40-ton tanks, or six 25-ton tanks, or 5 trucks, 4 carriers and 1 scout car, or 250 tons cargo	159'11''	31'	1'11'' for'd 5'0'' aft (light) 3'8'' for'd, 7'0'' aft (loaded) (tanks)	296 tons (light)	11 kts max. (Diesel) 10½ kts. max. (gasoline)	2,700 w/3 engines; 3,500 w/2 engines	Two 2-pdr. pompoms, or two 20 mm.	15-lb. DIHT to wheelhouse, foc'sle, shelter, DKM to deck and sides; 20-lb. gun shields; 2½'' plastic to bridge	12
LCT (3) Landing Craft, Tank (Mark 3)	Five 40-ton tanks or ten 3-ton trucks or 300 tons cargo	192'	31'	1'9'' for'd 4'3'' aft (light) 3'5'' for'd, 6'8'' aft (loaded) (tanks)	350 tons (light)	10.5 max. (loaded)	1,900 @ 10½ kts.; 2,700 @ 9 kts.	Two 2 pdr. pompoms or two 20 mm.	15-lb. DIHT to wheelhouse, conning tower; 15 or 20-lb. splinter shields to guns; 2½'' plastic to compass platform	12

Type	Capacity	Length	Beam	Draft	Weight or Displ.	Speed (Knots)	Endurance (Miles)	Armament	Armor	Crew
CRAFT—Con.										
LCT (4) Landing Craft, Tank (Mark 4)	Six 40-ton or nine 30-ton tanks or 12 loaded 3-ton trucks or 350 tons cargo	187′3″	38′8″	3′1″ for'd 4′3″ aft (loaded) (tanks) 1′0 for'd, 3′10″ aft (light)	200 tons (light)	8 max. (cont.)	500 @ 9½ kts.; 1,100 @ 8 kts.	Two 20 mm.	15-lb. DIHT to steering position, compass platform, gun positions, for'd winches, and 10 lb. to anchor reels	12
LCT (5) Landing Craft, Tank (Mark 5)	Five 30-ton or four 40-ton or three 50-ton tanks or 9 trucks or 150 tons cargo	117′6″	32′	2′10″ for'd 4′2″ aft (landing) 1′6″ for'd 3′9″ aft (light)	286 tons (landing) 134 (light)	8 max.	700 @ 7 kts. (loaded)	Two 20 mm.	2½″ plastic to pilot-house sides, 2″ gun position	13
LCT (6), Landing Craft, Tank—(Mark 6)	Four medium or three 50-ton tanks or 150 tons cargo; accommodations for 8 troops	120′4″	32′	3′4″ for'd 4′0″ aft (landing)	143 tons (light) 284 tons (landing)	8 max.	700 @ 7 kts.	Two 20 mm.	20-lb. STS on pilot-house (for'd), 15-lb. STS on pilothouse sides, 10-lb. STS splinter shields	12
LCT (8), Landing Craft, Tank—(Mark 8)	8 heavy tanks or 13 loaded 3-ton lorries or 350-tons cargo, 48 troops	226′	38′			12 (loaded)	2,500 @ 10 kts.	Two 20 mm.	To steering shelter	12
LCT (9), Landing Craft, Tank (Mark 9)	(No data available)									
LCT (A), Landing Craft, Tank (Armored)	Converted LCT (5); no further data available									
LCT (R), Landing Craft, Tank (Rocket)	Converted from LCT (2), (3)	159′11″ 192′	31′	4′3″ for'd 7′6″ aft		11.5 max.	1,400 @ 11½ kts. 2,700 @ 9 kts.	792–1,064 5″ rockets; two 20 mm.	Blast screen and wooden shelter to bridge	

U. S. AND BRITISH LANDING CRAFT

Type	Capacity	Length	Beam	Draft	Weight or Displ.	Speed (Knots)	Endurance (Miles)	Armament	Armor	Crew
CRAFT—Cont.										
LCV, Landing Craft, Vehicle	One 1-ton truck or 36 troops or 10,000 lb. cargo or 1 Bren carrier	36'3''	10'10''	2'2'' for'd 2'10'' aft (loaded) 1'6'' for'd 2'6'' aft (light)	11 tons (loaded) 7 tons (light)	9 (loaded)	68 @ 9 kts. 120 @ 7 kts. 200 @ max.		None	
LCV (2), Landing Craft, Vehicle—(Mark 2)	(No data available)									
LCVP, Landing Craft, Vehicle, Personnel	36 troops or 8,100 lb. cargo or 3 tons vehicles	36'	10'5¼''	3' aft 2'2'' for'd (light)	9 tons (light)	9 kts.	102 mi.	Two .30 cal. M. G.	¼'' STS on ramp and hull sides	3
LCW, Landing Craft, Air-Propelled	35 troops or 5 tons cargo			1'6'' max.		10 kts. (loaded)	200 @ 10 kts. loaded		Portable bulkheads to engine casing	
SHIPS										
LSD Landing Ship, Dock	3 LCT (5), (6) ea. with 5 med. tanks, or 2 LCT (3), (4) ea. with 12 med. tanks, or 14 LCM (3) ea. with 1 med. tank, or 1,500 long tons cargo, or 41 LVT's, or 47 DUKW's troops: 22 off., 218 men	457'9''	72'	8'2½'' for'd 10'½'' aft (light) 10'8'' for'd 13'8'' aft (light service) 15'5½'' for'd 16'2'' aft (seagoing, loaded) 30'9½'' for'd 29'9½'' aft (ballasted) 8'1'' for'd 9'11'' aft (well)	7,930 (seagoing) 4,032 (light)	17 kts. designed (max.)	8,000 @ 15 kts.	U. S.—One 5''/38 cal. DP, two 40 mm. twin, two 40 mm. quad., sixteen 20 mm. singles British—One 3''/50 cal. DP, four 2 pdr. pompoms, 16 20 mm. singles	Splinter protection for gun emplacements and pilothouse sides	17 off.; 237 men, LC-6 off.; 30 men.

Type	Capacity	Length	Beam	Draft	Weight or Displ.	Speed (Knots)	Endurance (Miles)	Armament	Armor	Crew
SHIPS—Con.										
LSM Landing Ship, Medium	5 medium, or 3 heavy tanks (150 tons max. pay load, beaching) or 6 LVT's or 9 DUKW's Troops: 54	203'6"	34'	6'11¼" for'd 3'5" aft (landing) 4'5" for'd 7'11½" aft (full load)	900 tons (loaded) 741 tons (landing)	13.3	3,500 @ 12 kts. est.		10-lb. STS—splinter shield to 20 mm. guns and pilothouse sides	Crew: 52
LST (British) See British Combined Operations Ships										
LST (1) Landing Ship, Tank—(Class 1)	Thirteen 40-ton or twenty 25-ton tanks and 36 loaded 3-ton trucks (150 tons total) Troops: 193	390'0" bp.	49'	5' for'd 13' aft (beaching)	5,410 gross tons 3,616 landing	17 (max.)	8,000 @ 14 kts. est.	Twelve 20 mm.· two 4" mortars	No data available	Crew 162
LST, Landing Ship, Tank UK designation LST (2)	2,100 tons (oceangoing max.), (350 tons main deckload), 1,060 tons diesel oil	328'0"	50'	8'0" for'd 14'4" aft (seagoing) 3'1" for'd 9'6" aft (landing) 1'6" for'd 7'5" aft (light)	4,080 tons (full load) 2,160 (landing)	10.8 kts. (max.)	6,000 mi. radius @ 9 kts.	U. S.—One 3"/50 cal. DP (omitted when 40's available), one 40 mm. (7 auth.), six 20 mm. (12 auth.) British—One 12 pdr., six 20 mm., four FAM	15-lb. STS splinter protection	(2 davit) 7 off., 204 men; troops: 16 off., 147 men; (6 davit) 9 off., 220 men; troops: 14 off., 131 men
LST (3), Landing Ship, Tank—(Class 3)	5 LCA, 1 LCT(6) or 14 trucks on upper deck; twenty-seven 25-ton tanks, or fifteen 40-ton tanks on tank deck; 168 troops	345'10" (o. a.)	54'	4'7" for'd 11'6" aft (beaching)	4,820 tons (loaded) 3,065 tons (beaching)	13.5 kts. (beaching displacement)	8,000 @ 11 knots (loaded) 1,000 @ 10 knots (landing condition)	4 twin 20 mm. guns and two 20 mm. single	15-lb. DHT around bridge, pilothouse, and gun positions	104

U. S. AND BRITISH LANDING CRAFT

Type	Capacity	Length	Beam	Draft	Weight or Displ.	Speed (Knots)	Endurance (Miles)	Armament	Armor	Crew
SHIPS—Con.										
LST (4), Landing Ship, Tank—(Class 4)	(No data available)									
LSV, Landing Ship, Vehicle	19–21 LVT's (AP 106, 107 none) 29-44 DUKW's 800 Troops	451'-453'	60'	21' (light)		19.5				
APc, Costal Transport	Troops—2 off., 60 enlisted; or 17 ton cargo	103' (wl)	21'3"	11' (loaded)	238 tons	9.5 kts. (cruising)	3,095 @ 9.5 kts.	Four 20 mm.	Splinter shielding	20
APD, High-Speed Transport (Destroyer)	One Marine rifle company; 4 LCP(L) or 4 LCP(R)	314'4"	31'8"	11'11"(max.) plus 4-foot sound dome	1,060–1,090 tons, 1,700 (loaded)	23 kts. (max.)	2,800 @ 20 kts. (approx.)	Four 3".50 cal. and five 20 mm. depth charges	Splinter shielding	112
APD (DE conversion)	4 LCVP, 162 troops, six ¼ ton trucks, two 1 ton trucks, 4 carts T4El, 4 Howitzers	306'	37'	14'9" (full load) 16'1" (with sound dome)	2,043 tons (full load)	23 kts. (max.)	5,000@ 15 kts. 2,000@ 23 kts.	One 5" .38 cal., three 40 mm. twins, eight 20 mm.		212
PGM, Motor Gunboat		110'	18'	5'5" (mean) 6'9" (loaded)	108 tons		690 @ 19 kts. 23 @ 12 kts. 2,890 @ 10 kts.	One 3"/23 or one 40 mm. forward, one 40mm. aft, four 50 cal. twins, one 60 mm. Army mortar, 1 multibarrel Bazooka	¼" plating to gun shields and control station	
VEHICLES										
LVT (1), Landing Vehicle, Tracked—(Mark 1)	4,500 lb. cargo or 20 equipped troops	21'6"	9'10"		16,900 lb. (unloaded)	4 kts. (water) 15 m.p.h. (land)	75 (land) 50 (water)	One .50 cal. M.G. One .30 cal. M.G.	None	3
LVT(2), Landing Vehicle, Tracked—(Mark 2)	6,500 lb. cargo or 24 equipped troops	26'1"	10'8"		25,200 lb. (unloaded)	5.4 kts. (water) 25 m.p.h. (land)	150 (land) 75 (water)	One .50 cal. M.G. One .30 cal. M.G.	Portable	3

Type	Capacity	Length	Beam	Draft	Weight or Displ.	Speed (Knots)	Endurance (Miles)	Armament	Armor	Crew
VEHICLES-Con.										
LVT(3) Landing Vehicle—Tracked (Mark 3)	8,000 lb. cargo or 24 equipped troops	24′1½″	10′10″		28,000 lb. (unloaded)	5.2 kts. (water) 25 m.p.h. (land)	150 (land) 75 (water)	One .50 cal. and one .30 cal. M.G.	Portable	3
LVT (4) Landing Vehicle, Tracked—(Mark 4)	6,500 lb. cargo max.	26′1″	10′8″		23,350 lb. (unloaded)	5.4 kts. (water) 25 m.p.h. (land)	150 (land) 75 (water)	One .50 cal. and one .30 cal. M.G.	Portable	3
LVT(A) (1), Landing Vehicle, Tracked (Armored) (Mark 1)	1,000 lb. cargo	26′1″	10′8″		25,200 lb.	5.4 kts. (water) 25 m.p.h. (land)	150 (land) 75 (water)	One 37 mm.; three .30 cal. M.G.	½″ plating on bow, cab, and turret; ¼″ plating elsewhere	6
LVT(A) (2), Landing Vehicle, Tracked (Armored) (Mark 2)	6,500 lb. cargo	26′1″	10′8″		27,600 lb.	5.4 kts. (water) 25 m.p.h. (land)	150 (land) 75(water)	One .50 cal. and one .30 cal. M.G.	½″ plating on bow, cab, and turret; ¼″ plating elsewhere	6
LVT(A) (4), Landing Vehicle, Tracked (Armored) (Mark 4)	2,000 lb. ammunition and gear	26′1″	10′8″		38,000 lb.	5.2 kts. (water) 25 m.p.h. (land)	150 (land) 75 (water)	One 75 mm. How.; one .50 cal. M.G.	½″ plating on bow and cab, 1″ on turret; ¼″ plating elsewhere	5
DUKW, 2½-Ton, 6 x 6 Amphibian Truck	25 equipped troops, or 12 loaded litters, or 5,000 lb. cargo	31′0″	8′0″		13,000 lb. (light) 18,600 lb. (loaded)	50 m.p.h. (land) 5.5 kts. (water)	400 @ 35 m.p.h. (land)	Varies	None	1
JEEP, ¼-Ton, 4 x 4 Amphibian Truck	800 lb. cargo	15′7″	5′4″		3,700 lb. (light)	50 m.p.h. (land) 4.7 kts. (water)	300 (land)	Varies	None	1
TERRAPIN (Mark 1)	5 tons (calm water) 4 tons (medium water) 50 men fully equipped	23′	8′9″		15,450 lb. (unloaded)	15 m.p.h. (max. land)				
TERRAPIN (Mark 2)	11,200 lb. (13,440 desired)	31′	8′6″		22,400 lb. (max.)	20 m.p.h. (land) 6–7 kts. (water)	125 mi. (land) 4 hours (water)			

BRITISH COMBINED OPERATIONS SHIPS

Additional material on these types will be forthcoming in Supplement No. 1.

TYPE, CLASS, SHIP	CAPACITY (Boats—Troops)	DIMENSIONS (Length o. a. x Beam)	TONNAGE (Gross)	MAXIMUM SPEED (Knots)	ENDURANCE (Miles—Knots)	ARMAMENT
LSI(S)—Landing Ship, Infantry (Small)						
Prince Charles Class Prince Charles Prinses Josephine Charlotte Prince Leopold Prinses Astrid	8—LCP(L) or LCA or LCS(M); 13–20 officers, 230–250 men; crew—32 officers, 175 men	360′ x 46′	3,647–3,730 (displacement)	22	600 @ 22 1,700 @ 10	Two 12 pdr. Two 2 pdr. Six 20 mm. Four M. G.
Prins Albert Class Prins Albert Prins Baudouin	8—LCP(L) or LCA or LCS(M); 39 officers, 340 men; crew—35 officers, 161 men	370′ x 46′	3,417–3,426 (displacement)	22	1,000 @ 20 1,700 @ 16 2,800 @ 12	Two 12 pdr. Two 2 prd. Six 20 mm. Four M. G.
LSI(M)—Landing Ship, Infantry (Medium)						
Queen Emma Class Queen Emma Princess Beatrix	6—LCA or LCS(M) and 2—LCM(1) on davits; 23 officers, 373 men; crew 28 officers, 187 men	380′ x 47′	3,976–3,988 (displacement)	22	3,000 @ 20 7,000 @ 13	Two 12 pdr. Two 2 pdr. Six 20 mm. Four .303″ (Pr. Beatrix 2)
Prince Henry Class Prince Henry Prince David	5—LCP(L) or LCA, and 1—LCS(M) and 2—LCM(2) or (3); 24 officers, 420 men	366′4″ x 57′1″	6,893	21.5	5,300 @ 20 7,500 @ 15 (est.)	Four 4″ DP Two 2 pdr. multiples Six 20 mm.
LSI(L)—Landing Ship, Infantry (Large)						
Westralia	Converting	431′1″ x 60′2″	8,108	15.5	9,100 @ 15 13,600 @ 12	Converting
Kanimbla	Converting	468′8″ x 66′3″	10,985	17	8,000 @ 15 10,500 @ 12	Converting
Manoora	8—LCA and 4—LCM(1)	463′5″ x 66′2″	10,856	16	7,000 @ 15 10,000 @ 12	Converting

TYPE, CLASS, SHIP	CAPACITY (Boats—Troops)	DIMENSIONS (Length o. a. x Beam)	TONNAGE (Gross)	MAXIMUM SPEED (Knots)	ENDURANCE (Miles—Knots)	ARMAMENT
LSI(L)—Landing Ship, Infantry (Large)—Con.						
Glenroy Class Glenroy Glenearn Glengyle	Glengyle— 12 LCA or LCS(M) and 3 LCM, 34 officers, 663 men; crew—29 officers, 256 men; Others— 13 LCA or LCS(M) and 3 LCM, 65 officers, 1,033 men; crew—70 officers, 553 men; 20 LCA's to be fitted.	500′ x 66′4″	13,000–15,590	19	10,000 @ 18 12,000 @ 14.5	Glengyle— Two 2 pdr. multiples Twelve 20 mm. Four 2 pdr. Eight 20 mm. Glenearn— Six 4″ twin additional
Keren	9—LCA, 2—LCM, 1—LCS(M), 2—LCP(L), and 10—LCP(S); 70 officers, 1,430 men; crew—38 officers, 259 men	471′ (b. p.) x 64′	14,500	17	6,000 @ 17 10,000 @ 12	One 6″ One 3″ DP Ten 20 mm.
LSI(H)—Landing Ship, Infantry (Hand-Hoist)						
Brigadier	6—LCA and military personnel; crew—13 officers, 106 men	305′ x 39′	2,185	21	300 @ 22 500 @ 16	One 12 pdr. Four 3″ Marlin Six 20 mm.
Duke of Wellington	8—LCA and 2—LCA on deck plus military personnel; crew—13 officers, 106 men	349′ x 52′	4,160	20.5	320 @ 19 430 @ 16	One 12 pdr. Ten 20 mm. Eight .303 4FAM, 4PAC
Invicta	6—LCA and military personnel; crew—16 officers, 130 men	352′ x 50′	3,210	22	300 @ 22 450 @ 17	One 12 pdr. Four 20 mm. 4 FAM Four .303

BRITISH COMBINED OPERATIONS SHIPS

Additional material on these types will be forthcoming in Supplement No. 1.

TYPE, CLASS, SHIP	CAPACITY (Boats—Troops)	DIMENSIONS (Length o. a. x Beam)	TONNAGE (Gross)	MAXIMUM SPEED (Knots)	ENDURANCE (Miles—Knots)	ARMAMENT
LSI(H)—Landing Ship, Infantry (Hand-Hoist)—Con. St. Helier	6—LCA and military personnel; crew—13 officers, 106 men	291′3″ x 40′	2,580	17	750 @ 17 900 @ 14	Four 20 mm. Eight .303 Lewis guns
Royal Scotsman Class Royal Scotsman Royal Ulsterman	4–6 Bren carriers, 90–150 T. cargo, 6—LCA and 30–40 officers, 450 men (85 off., 780 men on short 4-day trips); crew—14 officers, 70 men	340′ x 47′6″	4,215–4,217	17	2,500 @ 17 4,000 @ 10	Two 2 pdr. Eight 20 mm.
Ulster Monarch	6—LCA, 150 T. cargo, and 60 officers, 520 men; crew—30 officers, 100 men	357′ x 46′2″	4,734	17	2,500 @ 17 3,000 @ 14 4,000 @ 10	One 12 pdr. Two 2 pdr. Four 20 mm.
LSS—Landing Ship (Stern-Chute) Princess Iris Class Princess Iris Daffodil	13—LCM(1) or 9—LCM(3), floated over stern chute (but not to exceed 16 T. each)	350′ x 58′6″	2,500	11.5	1,265 @ 11.5 1,656 @ 9	Four 2 pdr. Five 20 mm. Five Lewis
LSH—Landing Ship (Headquarters) Bulolo	4—LCP(L), 19 officers, and 243 men (crew), and 23 officers and 64 men (military); staff—13 officers, 50 men	412′ o. a. x 58′	6,266	16	7,700 @ 15 9,300 @ 12	One 6″ One 3″ AA Two 2 pdr. multiples Eleven 20 mm.
Largs	Crew—23 officers, 284 men	339′6″ x 51′4″	4,500	17	14,500 @ 12	One 6″ One 3″ AA Two 2 pdr. multiples Ten 20 mm.
Hilary	6—LCP(L)	420′ x 56′	7,400	14¼	5,250 @ 12 4,250 @ 14	
LSB—Landing Ship, Bombardment	(No data available)					

TYPE, CLASS, SHIP	CAPACITY (Boats—Troops)	DIMENSIONS (Length o. a. x Beam)	TONNAGE (Gross)	MAXIMUM SPEED (Knots)	ENDURANCE (Miles—Knots)	ARMAMENT
LSC—Landing Ship, Carrier Empire Elaine Empire Charmain	21—LCM; Troops: 15 officers, 280 enlisted men	416' x 67' x 26'-7'' (mean)		10 Cruising	2,300	
LSF—Landing Ship, Fighter Direction	(No data available)					
LSG—Landing Ship (Gantry) Dale Class Dewdale Derwentdale Ennerdale	15—LCM(1); may be lowered with 10 T. load; 15 officers, 200 men, total accommodated	460' (b. p.) 59'3'' beam Ennerdale: 462'6'' (b. p.) 59'3''	16,750 (loaded) 5,450 (light)	12.5	13,000 @ 12.5 max. 10,000 @ 11 kts. plus 5,000 with auxiliary	Four 20 mm. (Dewdale) One 4''.7 One 12 pdr. Three 2 pdr. Two .303'' (Ennerdale) (Derwentdale has two 20 mm.)
LST—Landing Ship, Tank— Bachaquero Class	2—LCM (or smaller) and 18 Churchills or 22-25 T. tanks or equivalent motor vehicles; (6' bow ramp can support 40 T. tanks); 12 officers, 195 men; crew—13 officers, 70 men	382'6'' x 64' Mean draft—10' Landing—4'0'' for'd, 15'0'' aft	4,890	9-10	Bachaquero—6,200 @ 10 Misoa—6,600 @ 10	Three 2 pdr. Six 20 mm. Two 4'' smoke mortars Two Lewis
Tasajera Class	2—LCM (or smaller) and 22-25 T. tanks or equivalent motor vehicles; (6' bow ramp can support 40 T. tanks); 12 officers, 195 men; crew—13 officers, 70 men	365' x 60'2'' 10' Mean draft	3,952	9-10		Three 2 pdr. Six 20 mm. Two 4'' smoke mortars Two Lewis

PERSONNEL LANDING CRAFT

LCP(1)	Landing Craft, Personnel	(Mark 1)
LCP(2)		(Mark 2)
LCP(N)		(Nested)
LCP(U)		(Utility)
LCW	Landing Craft	(Air-Propelled)
LCP(S)	Landing Craft, Personnel	(Small)
LCP(M)		(Medium)
*LCA	Landing Craft, Assault	
*LCP(L)	Landing Craft, Personnel	(Large)
*LCP(R)		(Ramped)
LCI(S)	Landing Craft, Infantry	(Small)
*LCI(L)		(Large)
LCR(S)	Landing Craft, Rubber	(Small)
LCR(L)		(Large)

*Major operational types.

LANDING CRAFT, PERSONNEL (MARK 1) LCP(1)

Operational use	Small wooden craft to be carried on British destroyers and towed by LCP(2).
Dimensions	20'0" x 6'0"; weight is 2,000 lb. light.
Capacity	18 men, 2 crew.
Speed	One can be towed @ 5.5 kts.; 2 @ 4.8 kts.

UK

LANDING CRAFT, PERSONNEL (MARK 2) LCP(2)

Powered landing craft carried on destroyers' davits.

Dimensions	25'6" x 6'3".
Weight	4,000 lb. (light).
Capacity	18 men, 3 crew.
Endurance	30 miles @ 10 kts. (max. loaded speed).
Propulsion	One 65-hp. Ford V-8 engine.
Fuel	30 imperial gallons gasoline.

UK

Photo of the LCP(N). LCP(1) and (2) are still building in Levant.

LANDING CRAFT, PERSONNEL (NESTED) LCP)(N)

Operational use	Small personnel carriers designed to be stacked in nested arrangement on the deck of transports. Capable of surf and inshore operation.
Description	Constructed entirely of plastic skin and wooden ribs. Formerly designated LCP(P), Plastic. Only 12 experimental craft have been built.
Endurance	Limited in heavy surf because of motor flooding.
Speed	10 to 12 m. p. h. (designed). Dimensions Length, 32'0" o.a.;
Draft	10" loaded (designed). beam, 7'10".
Weight	1,400 lb. w/o engine. Capacity 13 troops, 2 crew.
Propulsion	4-cylinder, 60-hp. portable outboard motor; 6 oars carried.

US

LANDING CRAFT, PERSONNEL (UTILITY) LCP(U)

(British Staff Requirements)

Operational use	To land forces, not under fire, in surf or on rocky coast; also to serve as general utility craft.
Description	Similar to the LCP(L); designed to be carried on davits or on decks.
Capacity	30 troops fully equipped.
Weight	10 tons (unloaded).
Draft	3'6" max. (aft).
Speed	15 kts. designed, 12 knots minimum top speed.
Endurance	150 miles @ max. speed (loaded).
Armament	None.
Crew	4.

UK

LANDING CRAFT, AIR PROPELLED LCW

(British Staff Requirements)

Operational use	To transport troops and equipment over shallow water, through swamps and other difficult water areas. British design for Far East operations.
Description	An airscrew-propelled shallow-draft craft fitted with a folding bow-ramp. These craft may be nested on deck or at davits.
Capacity	35 troops or 5 tons cargo.
Dimensions	Cargo space 31' x 9'. Draft 18" max.
Speed	10 knots (loaded).
Endurance	200 miles @ 10 knots (loaded).
Armament	None.
Armor	Portable bulkheads to engine casing.

UK

LCP(S) LANDING CRAFT, PERSONNEL (SMALL)

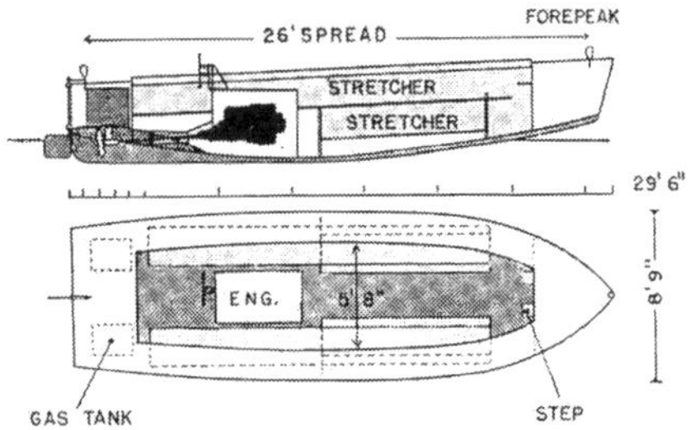

Operational use	For raids or second-echelon troop landings. Debarkation is by portable bow ladders shown secured on deck in photos. Some units have been equipped as water ambulances (drawing), carry 6 litters instead of benches. In all other craft 4 litters can be carried.
Description	Small wooden craft transported on ships' davits or on deck.
Capacity	20 fully equipped, 25 unequipped men; or 6 litters and 10 walking wounded.
Endurance	120 miles @ 8 kts. Speed, 8 kts. max. (loaded). @ 2,100 r.p.m.
Dimensions	Length, 29'6" o. a. Beam, 8'9".
Displacement	Light, 3 tons. Loaded, 5½ tons. Draft, 1'4" (mean) to 1'9" (mean).
Armament	One .303 Lewis gun. Crew, three.
Propulsion	One 60 or 65-hp. Chrysler engine; single screw.
Fuel	100 imperial gallons gasoline.

UK

Photo shows troop benches, propeller tunnel. ▶

Note single skin plank construction.

LANDING CRAFT, PERSONNEL (MEDIUM) LCP(M)

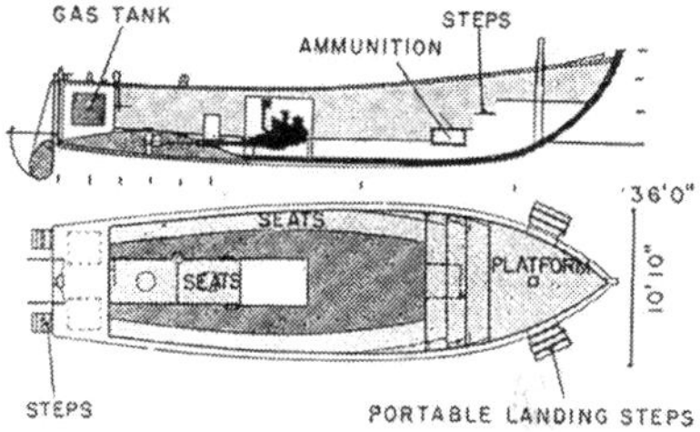

Operational use	For operations against rocky coasts and similar terrain which restrict the use of other personnel carriers.
Description	Small wooden craft adapted from the standard work-launch design. Disembarkation is by 4 portable ladders over bow and stern.
Capacity	20 fully equipped troops.
Endurance	112 miles @ 7½ kts. Speed, 7½ kts. max. (loaded) @ 2,800 r.p.m.
Dimensions	Length, 38'11'' o. a. (rudder up); beam, 10'1''.
Displacement	Light, 4½ tons. Draft, 1'6'' for'd, 1'7'' aft. Loaded, 7¼ tons. 2'1'' for'd, 1'7'' aft.
Armament	One .303 cal. Lewis gun; protected ammunition box for'd.
Crew	One officer per 3 craft, 3 men.
Propulsion	One Scripps Ford V-8, 65-hp. engine; single screw.
Fuel	100 imperial gallons gasoline in 2 linatex tanks.

UK

LCA Variations in the ¼" steel doors and the bow gunners' cockpit are apparent in the photos. Note the general similarity of this design to the LCVP and LCP(R).

Notice canvas hold-cover and scaling ladders in illustrations.

LANDING CRAFT, ASSAULT LCA

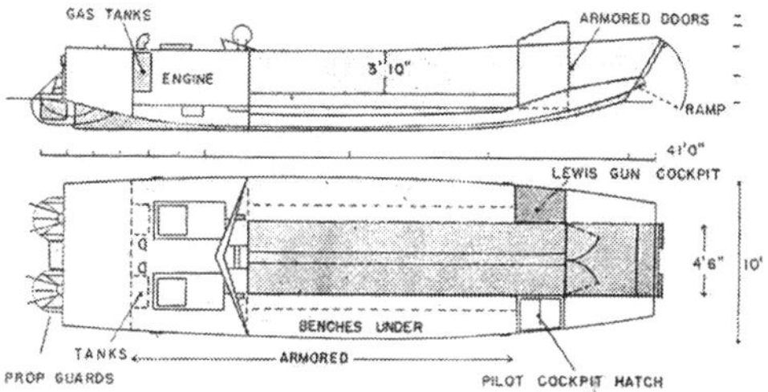

Operational use	For raiding operations or large-scale landings of troops from transports.
Description	The first British standard assault type, still being built in the Pacific Area. This type is standard equipment on most of their transports, usually carried on davits. The same design has been adapted to the LCS(M) and LCA(HR).
Capacity	35 troops and 800 lb. equipment. Troop space 18' x 8'10".
Endurance	50–80 miles @ 7 kts. (loaded), 65 miles @ full speed, 100 miles in special craft. Speed, 6 kts. (loaded), 10 kts. (light).
Dimensions	Length, 41'6" with prop. guards. Beam, 10'0".
Displacement	Light, 9 tons. Loaded, 13 tons. Draft, 1'1" for'd, 1'9" aft. 1'9" for'd, 2'3" aft.
Armament	1 Bren gun in port cockpit, 2–.303 Lewis guns, 2–2" mortars are being fitted aft.
Armor	10-lb. DIHT to bulkheads and hull sides, 7.8-lb. DIHT to hold decks (see drawing); 10-lb. parapets can be fitted.
Crew	1 officer per 3 craft, 4 men.
Propulsion	Two 65-hp. Ford V-8 engines; twin screws.
Fuel	64 imperial gallons gasoline; some units carry 98 gallons.

◀ This and upper left photo on opposite page are older type LCA.

UK

LCP (L)

Loaded LCP(L) beaching its troops. Debarkation has been improved by the addition of a bow ramp (LCP(R)).

▲▼ Both types shown during practice maneuvers.

LANDING CRAFT, PERSONNEL (LARGE) LCP(L)

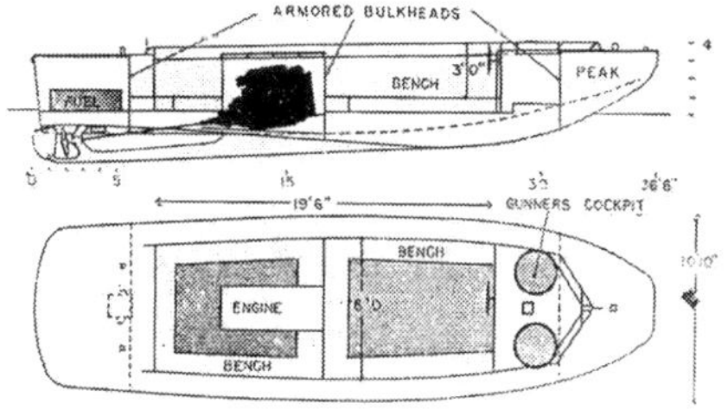

U. K. type ▼ U. S. type ▶

Operational use	Standard United Nations wooden assault boats for beaching operations.	
Description	Early United Kingdom type has enclosed cockpit as illustrated below, which has been used as a gunner's position in later U. S. model. Both types are being superseded in production by the LCP(R), LCVP. Both may be lowered from davits in a loaded condition unless otherwise specified by warning plate in boat.	
Nos. 1–500	U. K. type	Nos. 500–U. S. type.
Capacity	25 fully equipped troops, or 6,700–8,100 lb. cargo.	30–36 troops, or 6,700–8,100 lb. cargo, depending upon fuel load.
Speed	9–11 kts. (max.) (loaded).	8 kts. (fully loaded).
Endurance	120 miles @ full speed (loaded).	50 miles (gasoline). 130 miles (diesel).
Dimensions	Length, 36'8" o. a.	Beam, 10'10".
Displacement	Light, 13,000 lb.	Loaded, 18,000 lb.
Draft	2'6" aft. (light).	3'6" aft. (loaded).
Armament	One .303 Lewis gun.	Two .30 cal. in cockpit.
Armor	Nil. Some are being fitted with 10-lb. plating as shown on drawing below.	3 transverse 10 lb. bulkheads.
Crew	3 men, 1 officer/3 craft.	3
Propulsion	One Hall-Scott 250-hp. gas engine, or Kermath 225-hp. gas engine, or Gray 165–225-hp. Diesel. or Superior 150-hp. Diesel.	Same as U. K.
Fuel	200 gal. gasoline.	120 gal. gas or Diesel fuel.

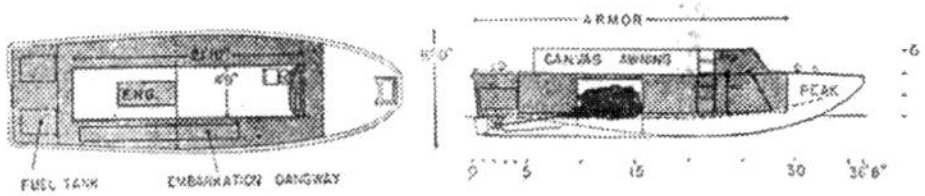

US—UK

LCP (R)

LANDING CRAFT, PERSONNEL (RAMP) LCP(R)

Operational Use—A personnel carrier similar to the LCP(L), with a ramp added to decrease debarkation time. This type is standard with the United Nations but is being superseded in production by the LCVP.

Description—Statistics are similar to the U. S. type LCP(L). Also like the LCP(L), these craft may be carried on APA, AKA, or APD davits from which they may be lowered when fully loaded. When transported on deck they can be lowered only in an unloaded condition by the slings.

Engines	Endurance (Miles Fully Loaded
225 hp. Gray Diesel	69
105 hp. Buda Diesel	145
115 hp. Chrysler Royal Gasoline	104
150 hp. Palmer Gasoline	80

US

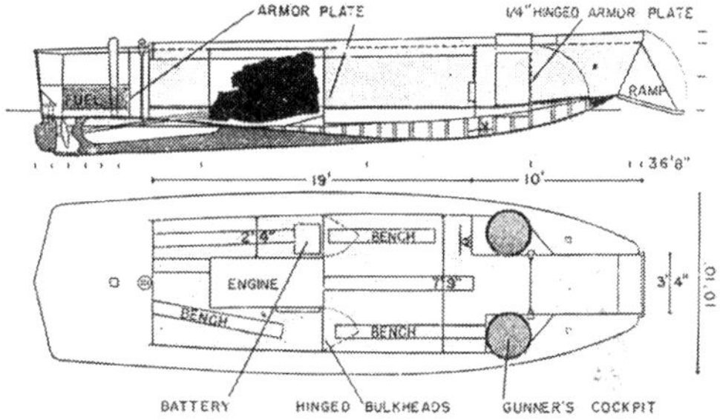

LCI (S)

Round CDL tower amidships is to be removed, as in the LCS(L).

LANDING CRAFT, INFANTRY (SMALL) LCI(S)

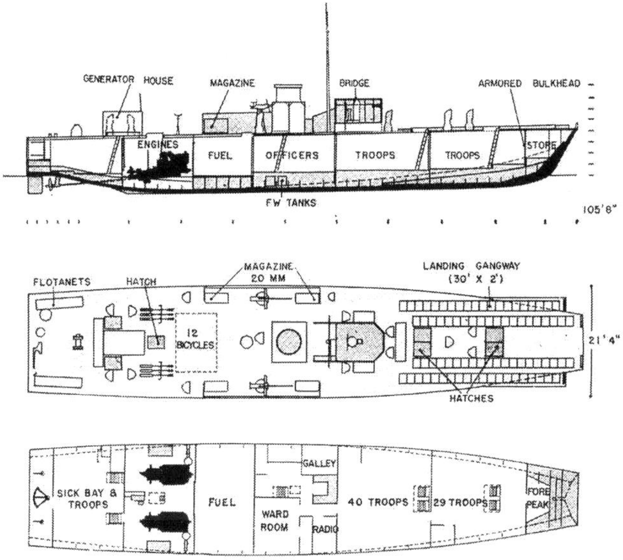

Troops are provided with hammocks in the compartments shown above.

Operational use	Medium-sized British infantry carrier designed for raider operations.
Description	Wooden hull and general design are the same as LCS(L), a large support craft. Troops debark over 4 bow ramps as in LCI(L).
Capacity	Six officers and 96 fully equipped troops below decks; 18 bicycles stored on main deck.
Endurance	550 miles @ 15 kts.; 700 @ $12\frac{1}{2}$ kts.
Speed	15 kts. max. permissible @ 2,000 r. p. m. $13\frac{1}{4}$ kts. max. continuous speed @ 1,600 r. p. m.
Dimensions	Length, 105'1'' o. a. Beam, 21'4''.
Displacement	Light, 63 tons. Loaded, 100 tons. Draft, 2'10'' for'd, 3'8'' aft. (loaded)
Armament	Two 20 mm. (2 or more to be fitted); two .303 Lewis guns.
Armor	10-lb. DIHT plating to deck, sides, gun positions, generator house, and forward bulkhead; $\frac{1}{4}$'' plate to bridge.
Crew	Two officers, 15 men, with accommodations.
Propulsion	Two Hall-Scott gasoline engines; 1,120 hp. or 1,500 hp. (supercharged); twin screws; silencers fitted.
Fuel	4,000 imperial gallons gasoline.

UK

20
20

196
196

LCI (L)

LANDING CRAFT, INFANTRY (LARGE)—Numbers 1-350 LCI(L)

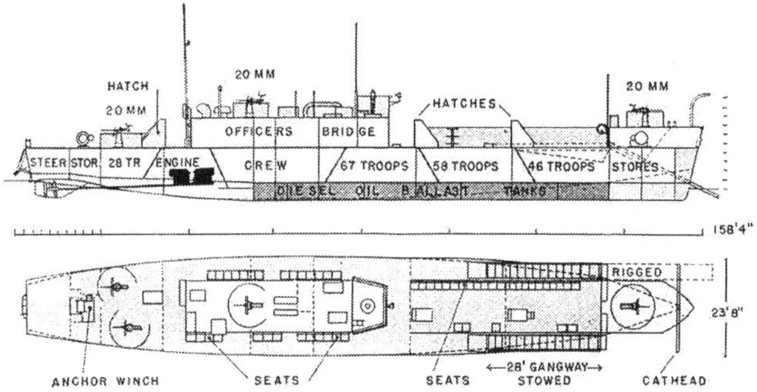

Operational use	An ocean going infantry carrier designed for direct unloading on the beach. Although living space, galley, and toilet are provided for accommodation of troops, the space is limited to a practical operational time of 48 hours.
Description	Early models, illustrated by the drawing, are undergoing the changes shown by the photos; heightened conning tower, mainmast moved forward.
Capacity	6 officers, 182 enlisted men, or 75 tons cargo.
Endurance	4,000 miles @ 12 knots; beaching draft allows for 500 miles @ 15 knots, or 1,500 miles @ 12 knots.
Speed	16 kts. max. (loaded) @ 650 r.p.m. 14 kts. max. (continuous).
Dimensions	Length, 158'5½" o. a. Beam, 23'3".
Displacement	Light, 216 tons. Draft 3'1½" mean. Landing, 234 tons. 2'8" for'd, 4'10" aft. Loaded, 380 tons. 5'4" for'd, 5'11" aft. Tons per inch immersion, 6.5 at landing draft, 7.0 at ocean draft.
Armament	Four 20 mm. (British also carry two .303 Lewis guns).
Armor	2" plastic splinter protection for guns, conning tower, and pilothouse sides.
Propulsion	2 sets of G. M. Diesels; max. B.H.P. 1,600; twin variable-pitch screws.
Fuel	130 tons fuel oil, 200 gallons lube. Crew, 3 officers, 21 men.

US

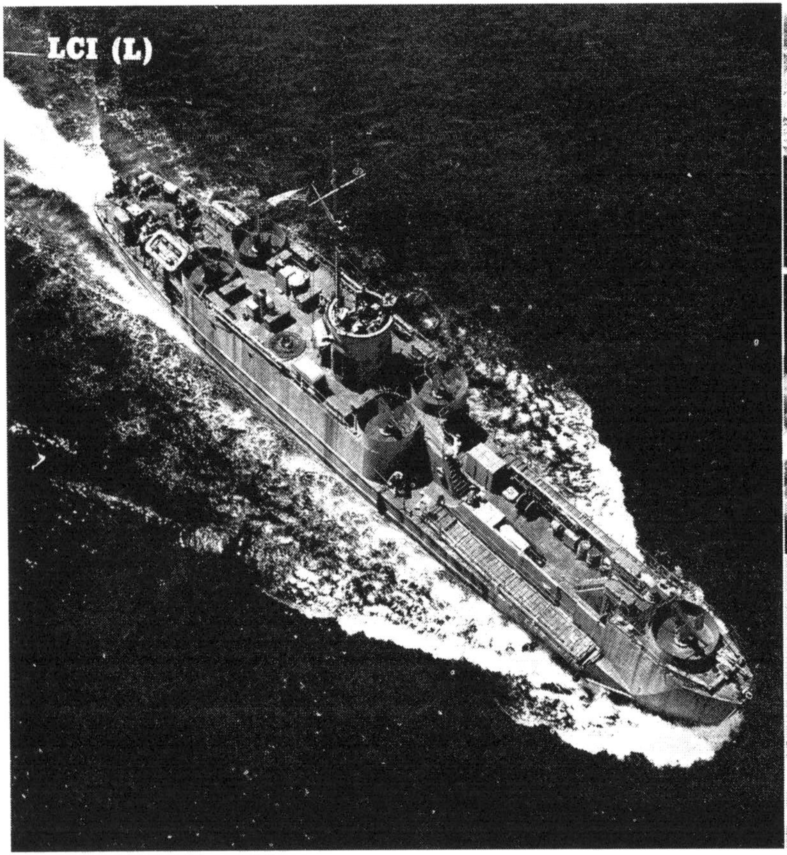
LCI (L)

LANDING CRAFT, INFANTRY (LARGE) Nos. 351— LCI(L)

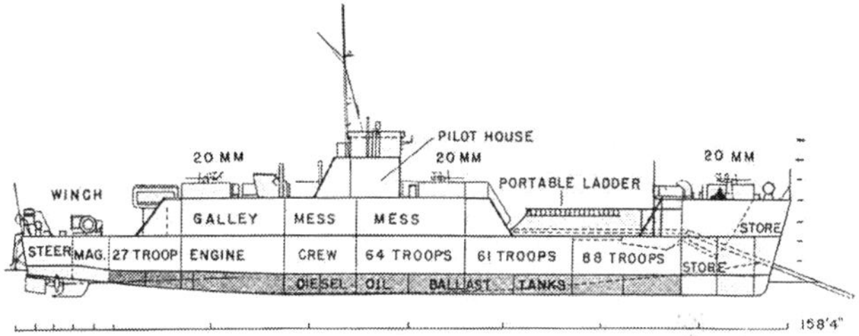

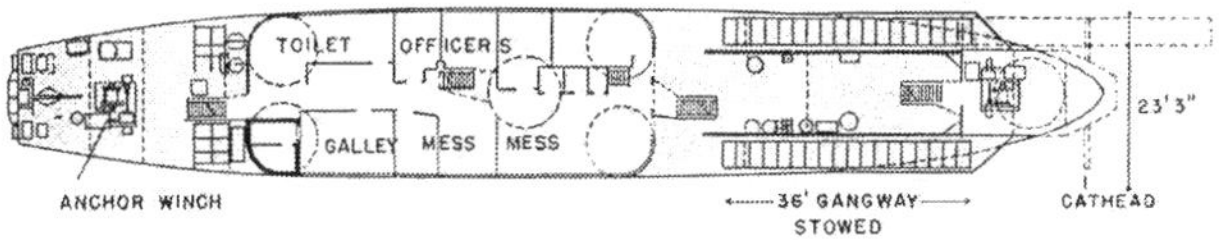

Dimensions and all statistics not mentioned below are the same as early model.

Capacity	9 officers, 196 enlisted troops; 32 tons cargo (oceangoing.)	
Endurance	8,000 miles @ 12 knots and 110 tons fuel.	
Displace-ment	Landing, 250 tons. Loaded, 387 tons.	Draft, 3'0" for'd, 5'0" aft. 5'8" for'd, 5'8" aft.
Armament	Five 20 mm.	
Armor	2" plastic on guns, conning and pilothouse sides.	
Crew	4 officers, 24 men.	
Fuel	110 tons fuel oil, 37 tons water, 5 tons lube.	

US

All production in this class after 1 June 1944 will incorporate centerline ramp as shown on left. Boilers and troop spaces have been rearranged but statistics are unchanged.

▲ A smaller type used by the Army.

▲ LCR(S) and their crews.

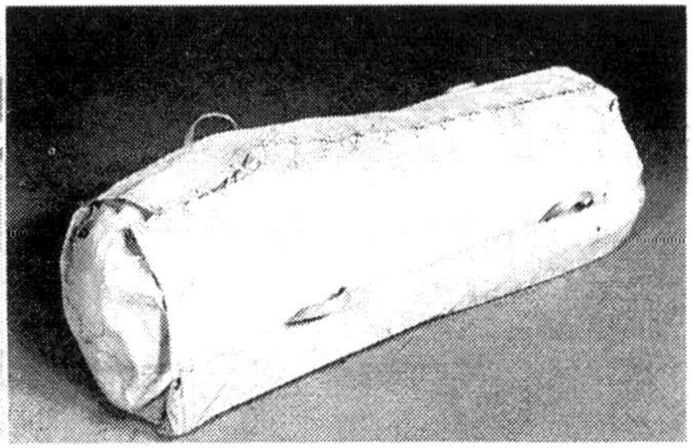

Each LCR(S) and equipment may be stowed in a foot kit. ▲

▲ LCP(L) unloading LCR's.

▲ LCR(S) equipped with outboard.

▼ LCR(L)'s in operation. A light machine gun may be mounted in bow position.

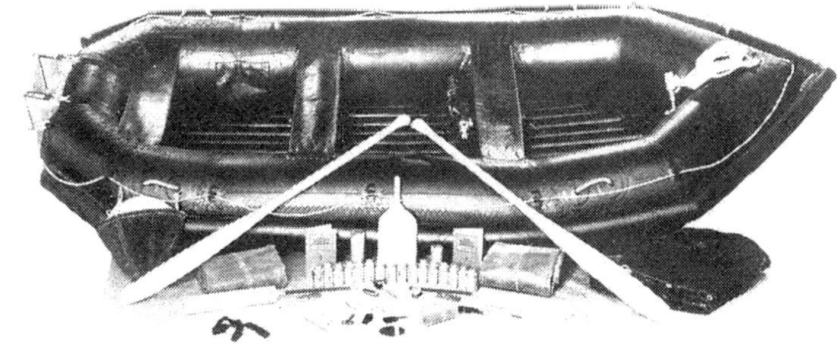

▲ LCR(S) ▼ LCR(L). Photos illustrate standard equipment for each type.

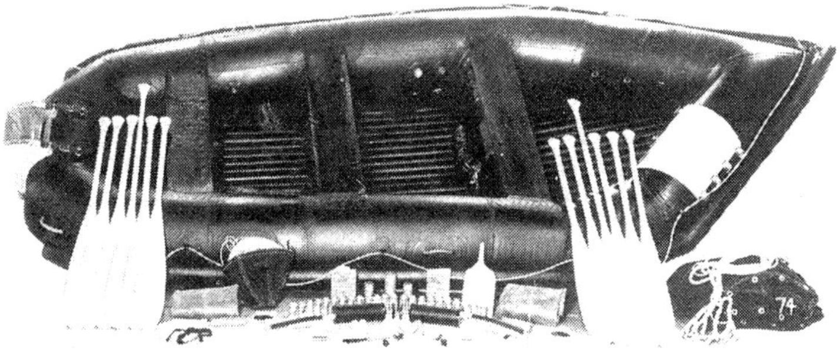

Operational Use	To land patrols and raiders secretly (usually by night) from APD's, submarines, or transports. May also be used by air-borne troops.
Description	Collapsible rubber rafts powered by a portable outboard motor or paddled to the beach.

SMALL 7-MAN RUBBER BOAT LCR(S)

Dimensions	12'5'' x 5'11'' beam.
Weight	210 lb. light, 277.4 lb. with motor.
Speed	3½–4½ knots with 6 hp. outboard motor.

US

LARGE 10-MAN RUBBER BOAT LCR(L)

Dimensions	16'0'' o. a. x 8'0'' beam.
Weight	395 lb. light, 474 lb. with motor.
Speed	3½–4½ knots with 9½ hp. outboard motor; 8 experienced paddlers can attain 66 yards per minute for 3 miles, but 57 yards per minute for 2 miles is normal. 2–3 miles paddling is maximum before fighting.
Armament	One .30 cal. M.G.
Towing	A 26' motor whaleboat can tow 8 of these craft in calm water, 7 in rough water, at 4–4½ kts.
Ferrying	Two LCR(L) can be carried in an LCVP or 6 in an LCM(3) for ferrying personnel and cargo.

US

VEHICLE AND TANK LANDING CRAFT

LBV(1)	Landing Barge, Vehicle (Mark 1)
LBV(2)	(Mark 2)
*LCV	Landing Craft, Vehicle
*LCVP	Landing Craft, Vehicle, Personnel
LCM(1)	Landing Craft, Mechanized (Mark 1)
LCM(2)	(Mark 2)
*LCM(3)	(Mark 3)
LCM(4)	(Mark 4)
LCM(5)	(Mark 5)
*LCM(6)	(Mark 6)
LCM(7)	(Mark 7)
*LCT(1)	Landing Craft, Tank (Mark 1)
*LCT(2)	(Mark 2)
*LCT(3)	(Mark 3)
*LCT(4)	(Mark 4)
*LCT(5)	(Mark 5)
*LCT(6)	(Mark 6)
LCT(8)	(Mark 8)
LCT(9)	(Mark 9)
Z-Lighter	

*Major operational types.

LANDING BARGE, VEHICLE—(MARK 1) (DUMB), MARK 2 (POWERED) **LBV**

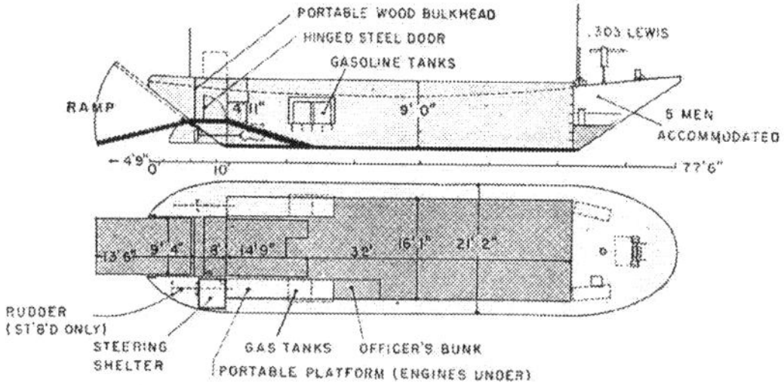

Operational use Commercial barges converted to transport and embark vehicles in the initial stages of the landing. Also used as ship-to-shore lighters for general cargo.

Description Steel "swim" barges of various dimensions, similar to the LBF and LBE. A 9'4" stern ramp services the hold. Mark 1 designates the "dumb" barges, Mark 2 the units with engines installed. All were formerly designated by size, such as LBV(S), etc. Drawing shows LBV(M).

Dimensions	Small	Medium	Large
Width of hold	13'–14'	14'–15'	15'–17'
Length of hold	46'	50'	60'
Over-all length/width	70'/18'	78'/20'	82'/23'
Height to underside deck	6'6"	7'6"	8'

Allow for 15% variations in any one class.

Capacity (Approximate) (S) 100 tons, (M) 150 tons, (L) 200 tons. Two 6-pdr. guns, 3 lorries (L and M) or two 3-ton lorries, one 15-cwt. lorry, two 30-cwt. lorries, (L and M); two 3-ton or three 15-cwt. lorries (S). 11 tons max. weight for ramp.

Speed (S), (M)—3–5 kts. (maximum) (loaded); (L)—6 kts (maximum) (loaded).

Endurance 360 miles @ 5 kts.

Draft (Average) 4'–4'6" mean (loaded).

Armor Concrete hull sides; 2½" plastic to gas tanks when fitted; 2 portable steel bulkheads inboard of ramp.

Propulsion Mark 1 dumb, Mark 2 has 2 Chrysler RM Gas Engines, one rudder.

UK

▼ Notice .50 cal. M.G. positions on bulwark.

▼ Note similarity between LCV (beached) and LCM(3) (tied up).

LANDING CRAFT, VEHICLE **LCV**

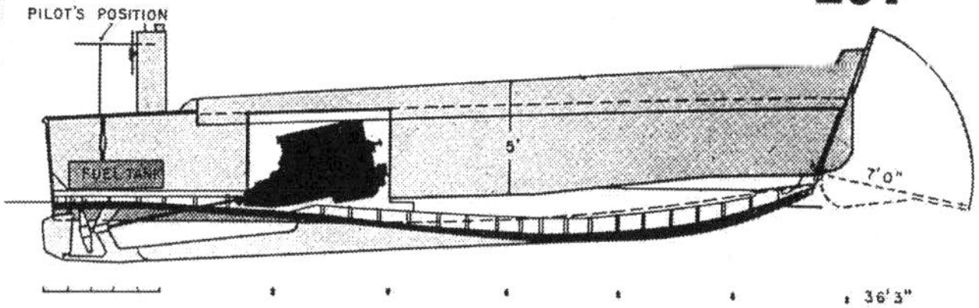

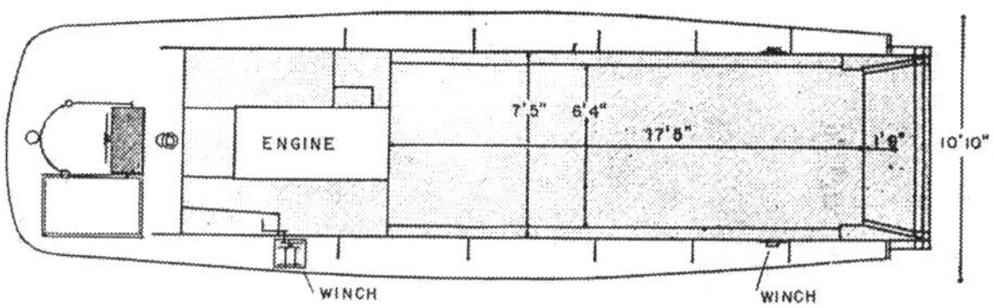

Operational use	Small vehicle carrier for initial and subsequent landing waves. Also used as personnel and cargo carrier.
Description	Small, wood-hulled craft similar to the LCM in design. It has recently been superseded in production by the LCVP, an improved armed adaptation of the same hull. Transported on deck (APA, AKA, LST), in hold (LSD, LCT), or on single- or 3-tiered davits. It is not designed to be lowered when loaded but may be hoisted with maximum of seven litters in calm sea.
Capacity	One 1-ton truck, or 36 troops, or 10,000 lb. cargo or 1 Bren carrier (tight).
Endurance	68 miles @ 9 kts.; 120 miles @ 7 kts.; 200 miles (max.) (later models).
Speed	9 kts. (loaded).
Dimensions	Length, 36'3" o.a. Beam, 10'10".
Displacement	Light, 7 tons. Draft, 1'6" for'd, 2'6" aft. Loaded, 11 tons. 2'2" for'd, 2'10" aft.
Armament	One .303 Lewis gun on British models only.
Crew	3 men, 1 officer per 3 craft. Armor, none.
Propulsion	Various Diesel or gasoline engines; 225–250 h.p.

US

LCVP

This type is the standard davit boat on APA's.

LANDING CRAFT, VEHICLE, PERSONNEL **LCVP**

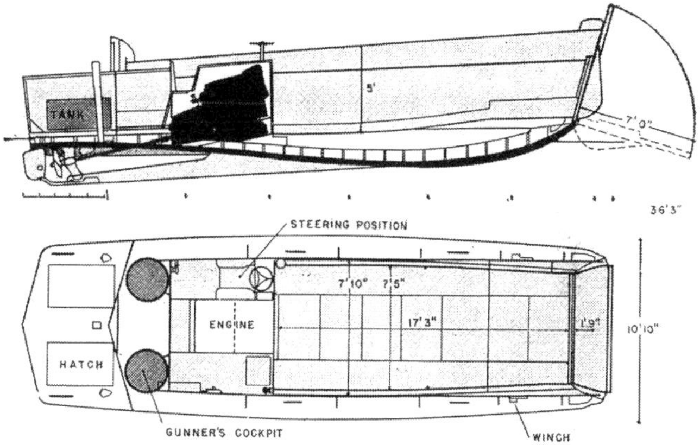

Description	Essentially an improved LCV with steering control and gunners' cockpits in the hold. This design is standard equipment for all ships and supersedes LCV in production.	
Capacity	36 troops or 6,000-lb. vehicle or 8,100 lb. general cargo, provided center of gravity is kept low. Can be lowered when loaded only from davits.	
Endurance	102 miles.	Speed, 9 kts.
Dimensions	Length 36′ o. a. Beam, 10′5¼″.	
Displacement	Light, 18,000 lb.	Draft, 3′ aft. 2′2″ for'd.
Armament	Two .30-cal. M. G.	Propulsion, 225-hp. Diesel. (Gray) or 250-hp. Gasoline (Hall-Scott)
Armor	¼″ STS, ramp and sides.	
Crew	3.	

US

LCM (1)

Note slight variations in steering shelter design.

▲ Portable ramp extensions for landing 3'' AA carriage.

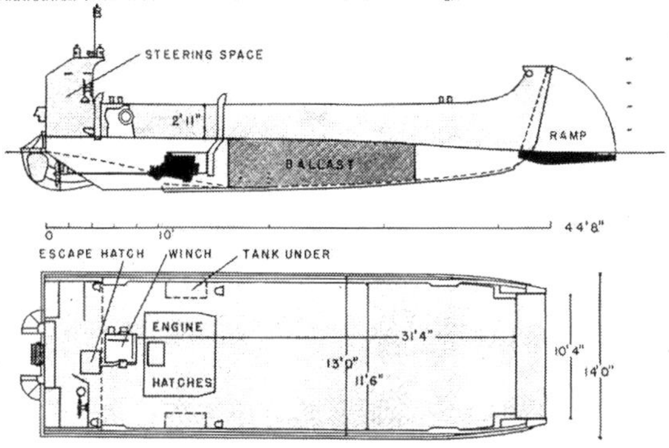

LANDING CRAFT, MECHANIZED—(MARK 1) **LCM (1)**

Operational use	To land a single tank or miscellaneous vehicles on a beach slope greater than 1/43. Cargo and personnel carrying are secondary functions.
Description	The first British small tank-landing craft, since superseded by LCM (3). Transported on decks of landing ships and lowered by booms, davits, stern chutes, or gantry cranes.
Capacity	Tanks or trucks of less than 16 tons total, or 100 men or mixed loads— One 25 pdr. gun and 1 DUKW or One DUKW and 1 kitchen trailer or Two 37 mm. AT guns and 2 weapon-carriers or One 37 mm. AT gun and carrier and 2 jeeps or 1 staff car or Six jeeps or Two 37 mm. AT guns and 4 jeeps. Some units can carry 17¼ tons, increasing draft 1''.
Endurance	56 miles @ 7½ kts. Speed, 7½ kts. (loaded) @ 2,000 r. p. m.
Dimensions	Length, 44'8'' o. a. Beam, 14'0''
Displacement	Light, 21 tons. Draft, 1'4'' for'd, 2'9'' aft. Loaded, 35 tons. 1'6'' for'd, 4' aft.
Armament	Two .303 Lewis guns.
Armor	8 or 10 lb. bulwarks, deck, control and fuel tanks.
Crew	1 officer per 3 craft, 6 men; no living accommodations on board.
Propulsion	Two 60 hp. Thornycroft or Chrysler engines; 28'' twin screw. This craft can be hoisted with a load of up to 10 tons in craft. Some units are fitted with davits slinging eyes 26'9'' apart.
Fuel	100 gals.

UK

LANDING CRAFT, MECHANIZED—(MARK 2) **LCM (2)**

Operational use	To land one light tank or motor vehicle (photo left).
Description	U. S. model, generally superseded by LCM (3).
Capacity	One 13½-ton tank or 30,000 lb. cargo or 100 men.
Dimensions	45' o. a. x 14' 1'' Weight, 58,000 lb. Draft, 3' (mean).
Endurance	75 miles @ 7½ kts. (max. speed).
Armament	Two .50 cal. M.G. Armor, ¼'' control station.
Crew	4.
Propulsion	Two 100 hp. Kermath gasoline engines (twin screws).

US

LCM(3)

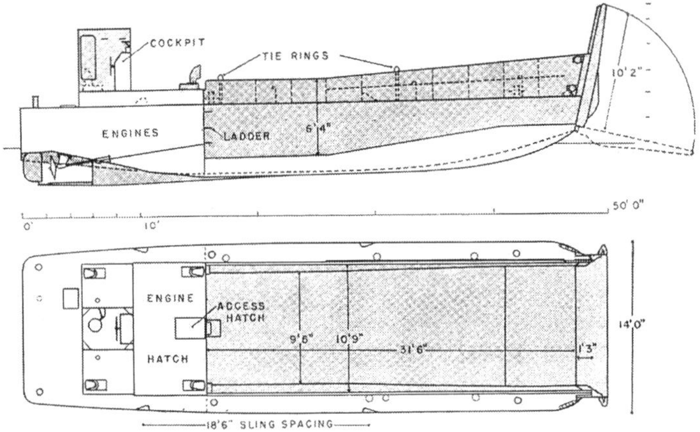

COCKPIT
TIE RINGS
10'2"
ENGINES
LADDER
6'4"
0'
10'
50' 0"
ENGINE
ACCESS HATCH
HATCH
9'6"
10'9"
31'6"
1'3"
14'0"
18'6" SLING SPACING

◀ **LANDING CRAFT, MECHANIZED (MARK 3)**

LCM(3)

Operational use	Designed to land one medium (30-ton) tank or motor vehicles directly on beach.
Description	Two designs are in use, the HIGGINS type constituting the majority. Recent changes include addition of machine guns and magnesyn compass, higher bulwark, and slight rearrangement aft.
Capacity	One 30-ton tank or 60,000 lb. of cargo, or 60 troops. BUREAU type (few in number) carries 120,000 lb. of cargo.
Endurance	850 miles @ 6¼ kts.; 500 miles @ 7¼ kts.; 140 miles @ full speed. Speed, 8 kts. (loaded); 11 kts. (normal load).
Dimensions	Length, 50′0″ o. a. Beam, 14′1″.
Displacement	Light, 52,000 lb. Draft, 3′0″ for'd, 4′0″ aft. Loaded, 52 tons.
Armament	Two .50-cal. M.G.
Armor	¼″ HTS sides to control station. British are fitting bullet-proof mattresses.
Crew	4.
Propulsion	Two 110-225 hp. Diesels of different designs; twin screws.

US

◀ LCM(3) during trials. In the foreground the BUREAU type debarks a 30-ton General Grant tank. In the background is a HIGGINS design.

LCM(4), (5) and (7) are British models. See table for staff requirements.

◀ **LANDING CRAFT, MECHANIZED (MARK 6)**

LCM(6)

A Higgins type LCM(3) with 6′ added to the hull amidships. Photo shows new portion unpainted. Other characteristics are similar to LCM(3).

US

Photos show canvas hold-covering spread on portable tubular rafters. Portable steel plates cover the bow.

▲▼ LCT(2) during trials. Other photos show LCT(1).

LCT (1) LANDING CRAFT, TANK (MARK 1)

CONFIDENTIAL

LANDING CRAFT, TANK (MARK 2) LCT (2)

Publication and Distribution Branch—Division of Naval Intelligence

Description	British prototype for all later large tank-landing craft, now considered obsolete and used principally for training.
Capacity	Three 40-ton, six 25-ton, or six 16-ton tanks, or 250 tons cargo.
Dimensions	Length, 152′ o. a. Beam, 29′. Draft, 3′ for'd (loaded) 5′9″ aft (loaded)
Displacement	Light, 226 tons.
Endurance	900 miles @ 10 kts. (max. speed)
Armament	Two 2 pdr's.
Propulsion	2 Hall-Scott gas engines; twin screws and rudders.

UK

▼LCT (1) Unloading.

Operational use	Large oceangoing landing craft designed to land tanks or vehicles directly on beaches sloping at least 1/32.
Description	Similar in appearance to other British types, the LCT (1), (2), (3), and (4).
Capacity	Three 40-ton or six 25-ton tanks, or 5 trucks, 4 carriers, and 1 scout car, or 250 tons of cargo.
Endurance	2,700 miles with 3 engines; 3,500 miles with 2 engines. Speed 11 kts. (max.) Diesel; 10½ kts. (max.) gasoline
Dimensions	Length 159′11″ o. a. Beam 31′0″
Displacement	Light, 296 tons; Loaded (trucks); Loaded (tanks). Draft 1′11″ for'd, 5′0″ aft.; 2′9″ for'd, 6′8″ aft.; 3′8″ for'd, 7′0″ aft.
Armament	Two 2-pdr. pompoms or two 20 mm.
Armor	15-lb. DIHT plating to wheelhouse, foc'sle shelter, DKM to deck and sides; 20 lb. splinter shielding to gun positions, 2½ plastic to bridge.
Crew	2 officers, 10 men, with accommodations; refrigerator provided in tropics.
Propulsion	3 Paxman Diesel (500 hp. each), or 3 Napier Lion gas engines (350 hp. each); 3 screws, 36″–40″, single rudder.

UK

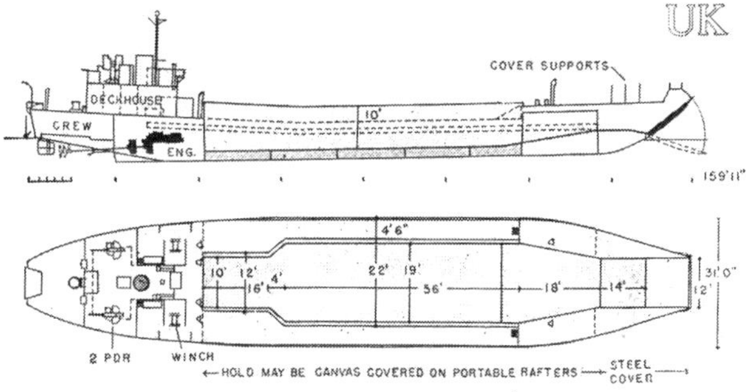

LCT (3)

Notice similarity to LCT (1), (2), and (4). Some units have been converted to LCF(3), LCG(3) and LCT(R).

ONI 226 · CONFIDENTIAL

Publication and Distribution Branch—Division of Naval Intelligence

LANDING CRAFT, TANK (MARK 3) LCT (3)

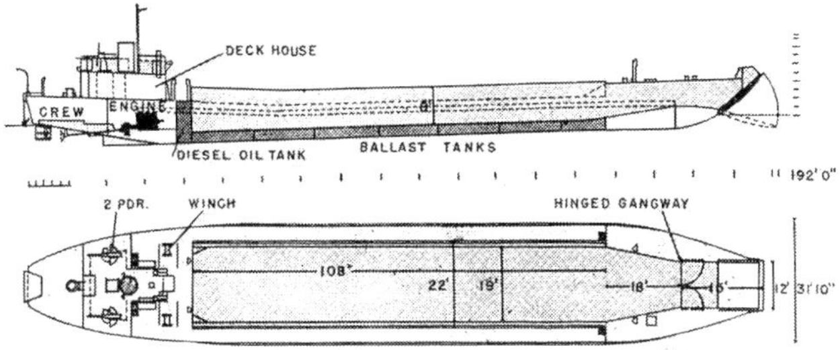

HOLD MAY BE COVERED WITH CANVAS ON PORTABLE RAFTER – 5' O.C.

Description	British type, longer than previous models. It can land tanks on slopes greater than 1/33, can only be transported on LSD.		
Capacity	Five 40-ton tanks or ten 3-ton trucks or 300 tons cargo.		
Endurance	1,900 miles @ 10½ kts.; 2,700 miles @ 9 kts.	Speed, 10½ kts. (max.) (loaded)	
Dimensions	Length 192'0" o. a. Beam 31'0".		
Displacement	Light, 350 tons.	Draft,	1'9" for'd, 5'3" aft.
	Loaded (tanks).		3'5" for'd, 6'8" aft.
	Loaded (trucks).		2'2" for'd, 6'0" aft.
Armament	Two 2-pdr. pompoms or two 20 mm.		
Armor	15-lb. DIHT over wheelhouse, to conning tower and anchor reels; 15- or 20-lb. splinter protection to gun positions; 2½" plastic protection to compass platform. Armored doors (gangways) in hold at bow.		
Crew	2 officers, 10 men, with accommodations.		
Propulsion	2 Paxman 500 h.p. Diesels; twin 40" screws and rudders.		
Fuel	24 tons oil stowed.		

UK

LCT (4)

LCT(4) differs from previous models in deckhouse arrangement aft. Note steel doors in hold at the bow. ▼

This craft has also been converted to LCF (4) and LCG (4).

LANDING CRAFT, TANK (MARK 4) LCT (4)

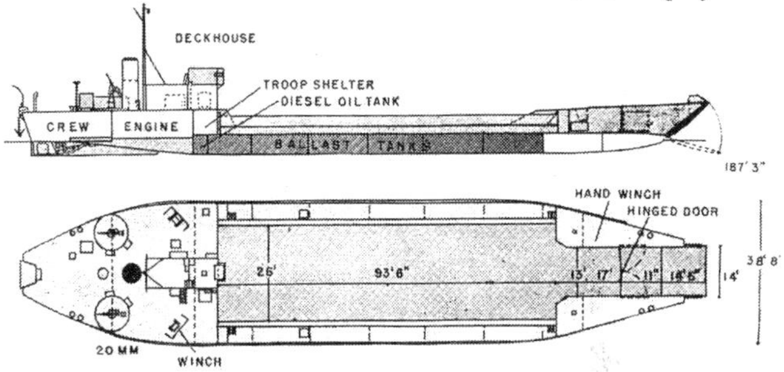

Opera-tional use	Besides the function of landing tanks and vehicles directly on beaches, these craft perform many varied base duties, such as net-tending (photo).		
Description	A British development of LCT(1)–(3) designs. Shorter and broader, this craft is capable of beaching on slopes greater than 1/150. Note that hold, unlike previous models, cannot be covered.		
Capacity	Six 40-ton or nine 30-ton tanks, or 12 loaded 3-ton trucks, or 350 tons of cargo.		
Endurance	500 miles @ 9½ kts.; 1,100 miles @ 8 kts.	Speed,	8 kts. (max. continuous) @ 1,375 r. p. m., or 5 kts. on single engine.
Dimensions	Length, 187'3" o. a. Beam, 38'8".		
Displace-ment	Light, 200 tons. Loaded (tanks). Loaded (trucks).	Draft,	1'0" for'd, 3'10" aft. 3'1" for'd, 4'3" aft. 1'10" for'd, 4'0" aft.
Armament	Two 20 mm., Army 40 mm. can be fired en route as AA protection.		
Armor	15-lb. plating to steering position, compass platform, gun positions, forward winches, and 10 lbs. to anchor reels.		
Crew	2 officers, 10 men, and tank crews, with accommodations.		
Propulsion	2 Paxman 500-hp. Diesels—twin 21" screws and rudders.		

UK

▼British units with portable ramp extensions, heightened deckhouse. Other views show recent 20 mm. additions on stern.

LANDING CRAFT, TANK (MARK 5) **LCT(5)**

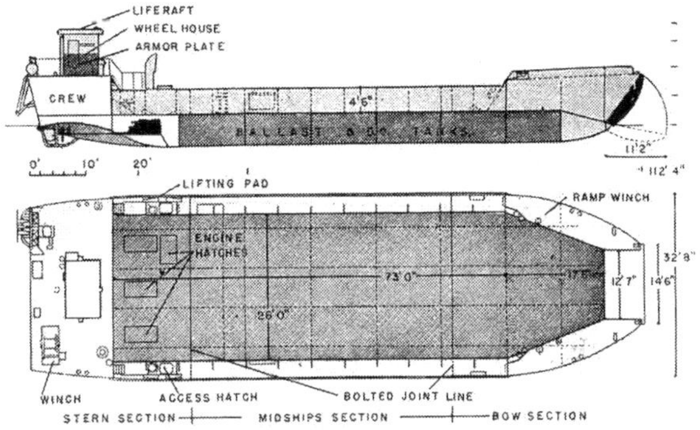

Description American design now standard for the United Nations, although superseded in production by LCT (6). This type is constructed in 3 watertight sections and may be transported disassembled on decks of larger vessels. An LST can carry 5 of these sections, or an assembled LCT, which is launched over the side by listing. An LSD can transport 3 loaded LCT (5)'s in its hold. Designed to beach on a slope of 1/120.

Capacity Five 30-ton, or four 40-ton, or three 50-ton tanks; or 9 trucks; or 150 tons of cargo. Freeboard limits load to 150 tons. No troop accommodations.

Endurance 700 miles @ 7 kts. (loaded).

Speed 8 kts. (max.).

Dimensions 117'6" o. a. Beam, 32'.

Displacement Light, 134 tons. Draft 1'6" for'd, 3'9" aft.
Landing, 286 tons. 2'10" for'd, 4'2" aft.
Tons/in. immersion, 7.5 in landing condition.

Armament Two 20 mm.

Armor 2½" plastic on pilothouse sides, 2" gun positions.

Crew 1 officer, 12 enlisted men.

Fuel 11.12 tons fuel, 140 gals. lube.

Propulsion 3 Gray 225-hp. Diesels; triple screw.

US

LCT (6)

Photo above shows sectional construction, typical also of the LCT(5).

▲ Unloading an Army 32-Ton Tank.

LANDING CRAFT, TANK (MARK 6) LCT (6)

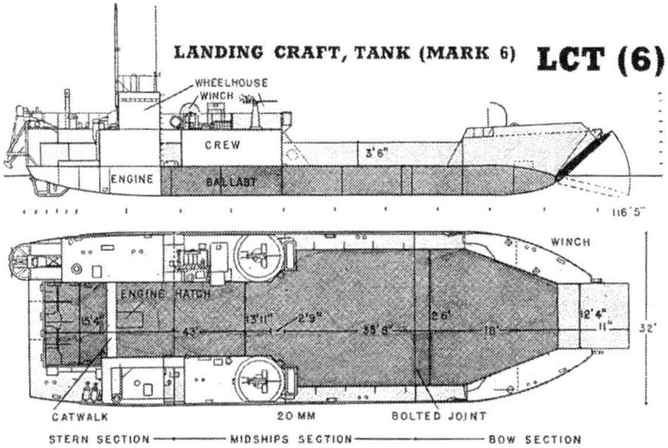

Operational use	Same as previous designs. In emergencies, this craft can serve as a floating bridge for unloading LST's. This is accomplished by mooring LST and LCT in line.
Description	LCT (5) hull modified to permit stern loading and increased living spaces.
Capacity	Four medium or three 50-ton tanks, or 150 tons cargo. Accommodations for 8 troops.
Endurance	700 miles @ 7 kts. Speed 8 kts. (max.).
Dimensions	Length, 120'4'' o. a. Beam 32'.
Displacement	Light, 143 tons; Landing, 284 tons. Tons/in. immersion, 7.5 in landing condition.
Draft	3'4'' for'd, 4'0'' aft (landing conditions).
Armament	Two 20 mm.
Armor	20-lb. STS on pilothouse for'd. 15-lb. STS on pilothouse sides. 10-lb. STS on splinter shields.
Crew	1 officer, 11 men.
Propulsion	3 Gray 225 hp. Diesels.
Fuel	11.12 tons fuel oil, 140 gal. lube oil.

US

LCT (8) LANDING CRAFT, (MARK 8)

(British Staff Requirements)

Operational use	To transport and land tanks and vehicles directly on beaches. Particularly fitted for operation in the Far East. Comparable to U. S. LSM. LCT(9) going into production, no statistics available.
Description	Larger and more seaworthy development of LCT(4). Capable of beaching in not more than 3 feet of water when fully loaded with motor transport or 4½ feet of water when fully loaded with heavy tanks. Slope of beach greater than 1/150.
Capacity	8 heavy tanks or 13 loaded 3-ton lorries or 350 tons cargo. Accommodations for 6 officers, 42 enlisted troops.
Dimensions	Length, 226′ o. a., Beam, 38′.
Speed	12 kts. (loaded).
Endurance	2,500 miles @ 10 kts.
Armament	Two-20-mm. guns.
Armor	To steering shelter.
Crew	2 officers, 10 men. Provisions for 7 days.

UK

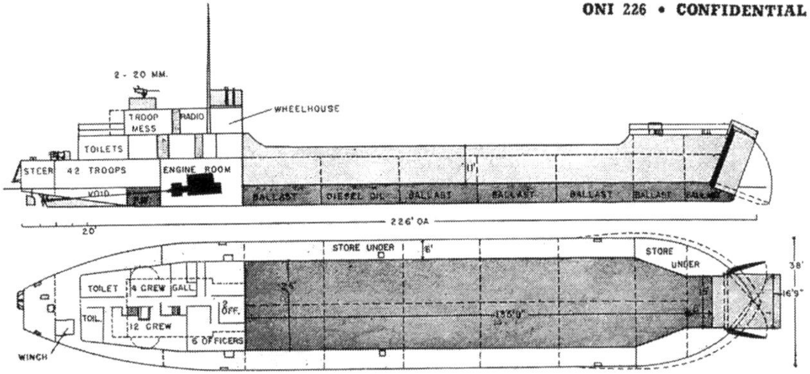

Z-LIGHTER

Operational use	A British harbor craft used in the Mediterranean area for ship-to-shore unloading of vehicles and general cargo.
Description	Similar to the British LCT (1)–(4) designs with a large reinforced bow ramp.
Capacity	Six 15-ton tanks or ten 6-wheeled lorries, or nine 4-wheel lorries, and 95 tons fresh water.
Dimensions	134′ o. a. x 30′.
Draft	2′2″ for'd, 4′0″ aft, 3′0″ mean with 80-ton load.
Endurance	236 miles.
Propulsion	2 Gray 100 hp. Diesels, twin screws. Newer engines are 120 hp. each.

UK

LANDING BARGE, FLAK **LBF**

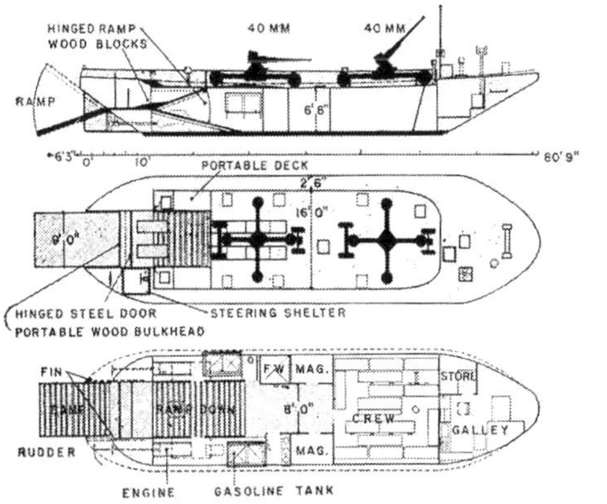

Operational use	To carry Army AA guns, providing an extended beach defense 3,000–5,000 yards seaward until the time when the guns may be landed.
Description	Similar in characteristics to the LBO, it is equipped with a 9' ramp like the LBV. All are steel-hulled, commercial "swim" barges.
Dimensions	60'–90' o.a. x 19'–24' x 5' mean draft.
Armament	Two 40-mm. (portable) Bofors. Two 20-mm. Hispano or 2 twin Lewis guns. LBG carries two 25 pdr.
Armor	3" concrete bottom and sides (4'6" up); plastic to gas tanks, steering station; steel deck over hold.
Crew	1 officer, 5 men; 1 officer, 15 troops accommodated.
Endurance	300 miles @ 5 kts. (max. continuous speed). Speed 6–7 kts. (max.)
Propulsion	Formerly dumb barges, now being fitted with two 150-h.p. Chrysler gas engines.
Fuel	600 gallons gasoline.

UK

LCS (S)(I) LANDING CRAFT, SUPPORT (SMALL) (MARK 1)

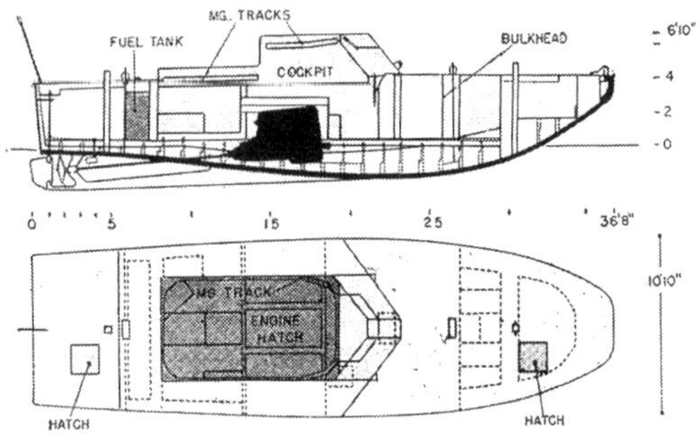

Operational use	To render close supporting automatic and rocket fire to assault waves; may furnish base of fire to facilitate initial maneuver of troops in a beach area. It is capable of beaching.
Description	Small wooden motorboat greatly resembling the LCC. Usually transported on the deck of APA's or AKA's, sometimes on davits.
Capacity	3–4 men plus crew and gunners.
Endurance	115 miles (full speed). Speed, 12 kts.
Dimensions	Length, 36'8" o.a. Beam, 10'10".
Displacement	Light, 20,000 lb. Draft, 3'6". 22,000 lb. (standard).
Armament	Two .50-cal. or three .30-cal. or one .50-cal. and two .30-cal. M.G.; 2 barrage rocket projectors carried on some units; 8 Mk. III smoke pots. When rockets are fitted, some guns are usually removed.
Armor	1/4" plating to bridge. Crew, 6. Fuel, 180 gallons.
Propulsion	1 Hall-Scott 250-hp. gas engine.

US

Photos Show Mark 1 With and Without Armament. Note Stern Smoke Pots.

LCS(S)(2)

LANDING CRAFT, SUPPORT (SMALL) (MARK 2)

All statistics, except those mentioned below, are identical with LCS(S)(1). Photos illustrate slight design differences.

Weight	23,000 lb.
Endurance	135 miles @ $11\frac{1}{2}$ kts. (max. speed).
Speed	10 kts.
Armament	One .50-cal. twin mount. Two .30-cal. M. G. 24 rockets are fired in 6 salvos. Two 12-rocket projectors and smoke racks.
Propulsion	1 Gray Diesel 225-hp.

US

LCS (M)(1) LANDING CRAFT, SUPPORT (MEDIUM) (MARK 1)

Operational use	To provide close support for assault waves with machine guns and smoke.
Description	Using the same hull design as LCA, this small wooden craft may be beached. Transported to the landing area on deck or davits and lowered in a loaded condition. Later models have a modified armament arrangement, as shown by the photo (bottom left).
Endurance	60 miles @ 10 kts. (loaded) Speed, 10 kts. (max.) @ 2,800 r.p.m.
Dimensions	Length, 41'2" o.a. Beam, 10'0".
Displacement	Light, 8.9 tons. Draft, 1'5" for'd, 1'9" aft. Loaded, 10.8 tons. 1'8" for'd, 1'11" aft.
Armament	Two .50 M.G.'s; two .303 Lewis guns; and one 4" smoke mortar or one 20-mm.
Armor	10-lb. plating on sides, parts of deck, and bulkheads; 20-lb. plating to conning tower. (See drawing).
Crew	1 officer, 3 men, 7 gun crew; no living accommodations.
Propulsion	2 Ford V8 65-hp. engines; twin 19" propellers.
Fuel	64 gallons gasoline.

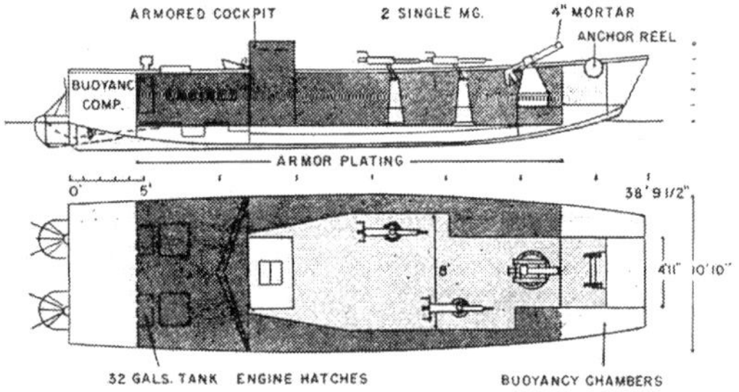

UK

LANDING CRAFT, SUPPORT (MEDIUM) (MARK 2)

LCS (M) (2)

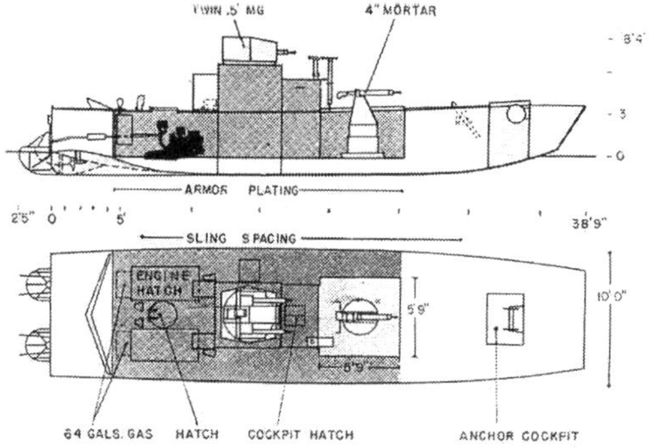

Description	Statistics for hull dimensions, armament, armor, engines, crew, etc., same as LCS(M) (1). Differences are illustrated by the drawing and the following figures.
Endurance	60 miles @ 9 kts.
Speed	9 kts. (max.) @ 2,800 r.p.m.
Displacement	Light, 10.5 tons. Loaded, 12.5 tons.
Draft	Light, 1'4" for'd, 2'2" aft. Loaded, 1'6" for'd, 2'5" aft.

UK

LCS (M)(3)

LANDING CRAFT, SUPPORT (MEDIUM) (MARK 3)

Description	An improved model of previous designs utilizing the same hull (with pointed bow), engines, and other equipment.
Endurance	90 mi. @ 9 kts.
Dimensions	Length, 41′2″ o.a. Beam, 10′1″.
Displace-ment	Light, 11.4 tons. Loaded, 12.8 tons.
Speed	9 kts. (max.).
Draft	1′6″ for'd, 2′0″ aft. light. 1′9″ for'd, 2′3″ aft. loaded
Armament	Two .50 twin power-operated Vickers M.G. in turret. Two .303 Lewis guns. One 4″ smoke mortar. 10 lb. H. E. bombs. Twelve Type 18 smoke generators.
Armor	10-lb. DIHT plating on hull sides and deck, bulkheads, and on conning tower and ammunition hoist.
Crew	1 officer, 3 men, 6 gun crew.
Propulsion	2 Ford V-8 Scripps 65 hp. engines.
Fuel	98 gals. gasoline.

UK

LANDING CRAFT, SUPPORT (LARGE) (MARK 1) LCS (L)(1)

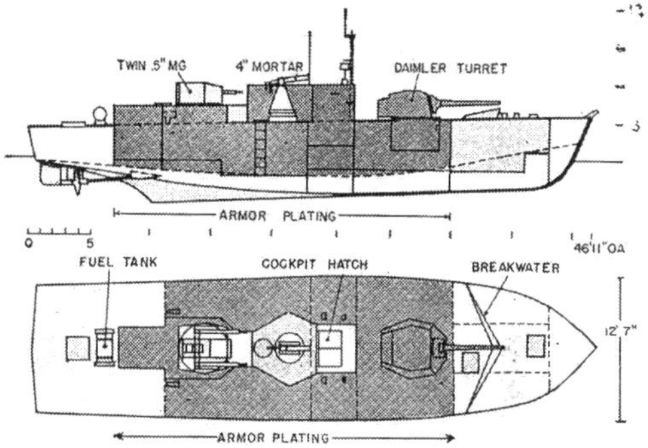

Operational use	To provide AA and surface support in the landing area. Not intended to be beached.
Description	A larger version of the wooden LCS (M) designs, with the addition of a powered turret. Usually transported on the deck of a ship, it may be lowered in a loaded condition. No living accommodations are provided for crew.
Capacity	Crew only. Crew, 1 officer, 4 men, 8 gunners.
Endurance	100 miles @10 kts. Speed, 10 kts. max. speed.
Dimensions	Length, 46'11" o. a. Beam, 12'7".
Displacement	Light, 20 tons. Draft, 2'2" for'd, 3'11" aft. Loaded, 24.5 tons. 2'4" for'd, 4'3" aft.
Armament	One 2-pdr. and BESA gun in Daimler Turret. One twin .50 cal. M. G., two .303 Lewis guns. One 4" smoke mortar.
Armor	10-lb. plating to deck and top of turret. 12-lb. plating to top of steering position. 15-lb. plating to hull sides, bulkheads, sides of turret. 20-lb. plating to sides of steering position. Drawing illustrates position of plating.
Propulsion	2 Gray Diesels @ 165 hp. each, twin 27" screws.
Fuel	146 imperial gallons oil.

UK

LCS (L)(2) LANDING CRAFT, SUPPORT (LARGE) (MARK 2)

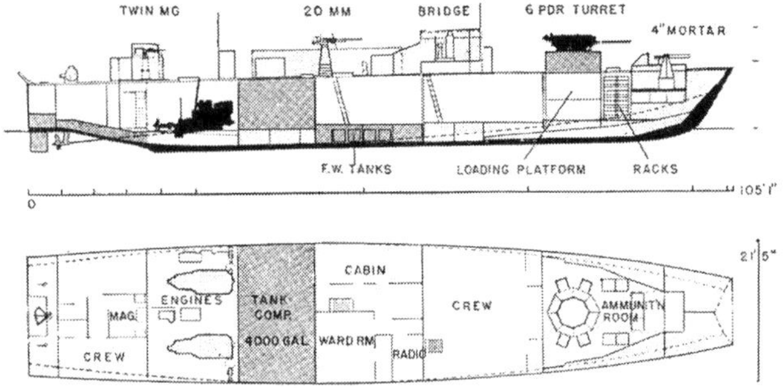

Operational use	To accompany landing craft in long-range operations to the beach area.
Description	Hull and many details are the same as LCI (S). Like LCS (L) (1), this craft is not intended to beach.
Endurance	700 miles @ 12½ kts. Speed, 14¾ kts.
Dimensions	Length, 105'1" o. a. Beam, 21'5".
Displacement	Light, 84 tons. Draft, 2'3" for'd, 3'0" aft. Loaded, 116 tons. 3'8" for'd, 3'4" aft.
Armament	One 6 pdr. in turret, two 20 mm. One twin .50 machine gun. One 4" mortar, two .303 Lewis guns.
Armor	10-lb. DIHT plating to deck, hull sides, and gun positions; ¼" plating on bridge.
Crew	2 officers, 23 men, with accommodations.
Propulsion	2 Hall-Scott 250 hp. gasoline engines (some are supercharged), twin 31" screws.
Fuel	4,000 imperial gallons.

UK

ONI 226 · CONFIDENTIAL
Publication and Distribution Branch—Division of Naval Intelligence

LCI GUNBOAT

Pending delivery of LCS(L) (3), some LCI (L)'s have been converted to gunboats in Pacific theaters as an interim expedient.
The armament of the different field conversions is as follows:

		40 mm.	20 mm.	.50 cal.	Rocket launchers
LCI (L)	77–82 345–348	2s.	3	6	10 Mk. 7
LCI (L)	365, 366 437–442	3s.	2	5	8 Mk. 1
LCI (L)	449–453 455, 457	2s.	4	6	10 Mk. 7 2 projectors Mk. 22
LCI (L)	372, 373, 454, 456 464–475, 458–463	3s.	4	?	10 Mk. 7
LCI (L)	21–24, 67–70	1s. & 1–3"/50 DP	4	6	? Mk. 7

All types have 2½" plastic armor to gun positions.
Photo below shows SOPAC conversion with a 3" mount.

US

LANDING CRAFT SUPPORT (LARGE) (MARK 3) LCS(L)(3)

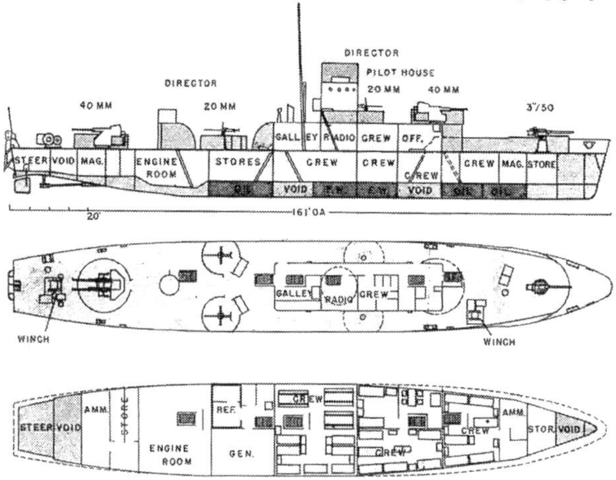

Operational use	To provide fire support for landing operations and to intercept and destroy inter-island barge traffic.
Description	60 units now being converted from LCI (L) hulls, but entirely rearranged internally, as shown by the drawings. This craft may be beached.
Armament	One 3"/.50 cal. DP., two twin 40 mm. MG., four 20 mm. MG.
Dimensions	Length, 158'5". Beam, 23'3".
Displacement	227 tons (light), 349 tons (normal), 383 tons (full load) (est.).
Draft	For'd 4'6", aft. 5' 10".
Armor	10 lb. splinter shields to gun positions.
Propulsion	2 sets of G. M. Diesel Engines, Model 6051, max. 1800 hp.
Fuel	85 tons Diesel oil.
Crew	5 officers, 68 enlisted men.

US

LCF (3) LANDING CRAFT, FLAK (MARK 3)

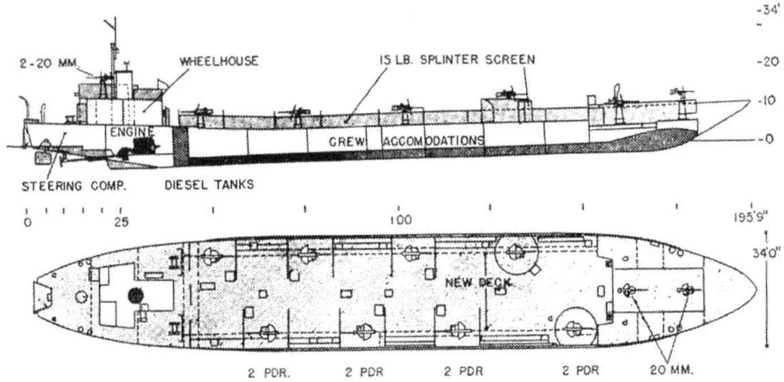

British units converted from LCT (3)

Dimensions	Length, 195'9"o.a. Beam, 31'0".
Displacement	440 tons (light).
Endurance	2,700 miles @ 9½ kts.
Speed	9½ kts. (max.) @ 1,100 r.p.m.
Draft	2'5" for'd, 6'9" aft., light.
Armament	Eight 2 pdr. pompoms and four 20 mm. (LCF 3–6); or four 2 pdr. pompoms and eight 20 mm. (LCF 7–18).
Armor	15-lb. DIHT plating to gun shields and gun positions; plastic on bridge.
Crew	2 officers, 10 men; gunners, 2 officers, 48 men, with accommodations.
Propulsion	2 Paxman 500-hp. Diesels, twin 40" screws.
Fuel	24 tons Diesel oil.

UK

LANDING CRAFT, FLAK (MARK 2) LCF (2)

Operational use	To protect assault forces against close-range air and surface attack prior to and during a landing. Also may engage concrete emplacements or M. G. posts by direct fire.
Description	Single unit, converted by the British from the LCT(2); all other conversions were from LCT(3) or (4) hulls. Drawing and photos illustrate new deck, armament, and hold arrangement for all types.
Endurance	2,000 miles @ 10 kts. Speed, 11½ kts. max. (loaded).
Dimensions	Length, 159'11" o. a. Beam, 31'0".
Displacement	455 tons. Draft, 3'6" for'd, 7'6" aft loaded.
Armament	Two twin 4" DP; three 20 mm.
Armor	20-lb. DIHT plating or 2½" plastic to bridge and gun deck.
Crew	2 officers, 11 men, 68 gun crew.
Propulsion	3 Paxman 500 hp. Diesels; triple screws. Fuel, 40 tons Diesel oil.

UK

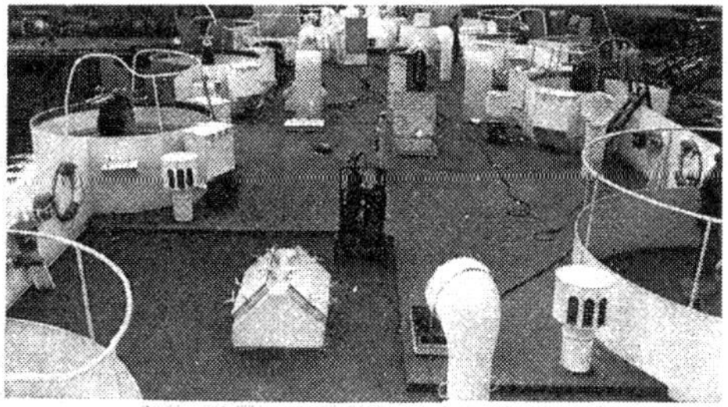

▲◀ LANDING CRAFT, FLAK (MARK 4) LCF (4)

British conversions from LCT(4).

Armament	Four 2 pdrs., eight 20 mm.
Armor	Probably same as LCF(3).
Endurance	1100 miles @ 8 kts.
Speed	9½ kts. max. (loaded).
Crew	2 officers, 10 men, 54 gun crew.
Propulsion	2 Paxman 500-hp. Diesels; twin 36" screws.
Fuel	12 tons Diesel oil.

UK

LCT (R)

LANDING CRAFT, TANK (ROCKET) LCT(R)

Operational use	To deliver a large volume of preparatory rocket fire on landing beaches.
Description	British conversions from LCT Marks (2) and (3.) False deck and rocket stands are readily removable so that the craft may continue as a tank-landing unit, once the initial landing is accomplished. Electrically fired in 24 salvos, the rockets cover an area of 750 x 160 yards, with a density of one rocket per 100 sq. yards.
Armament	792 (LCT (2)) or 1,064 (LCT (3)), 5'' rockets in stands of six @ 45° fixed elevation, 3,500-yard range. Rockets are fitted with (a) 29-lb. HE head with 7-lb. burster; (b) CSA smoke head; (c) incendiary head (for ranging at night). Two 20 mm.
Endurance	2,700 miles @ 9 kts. Speed, 11½ kts. (max.). 1,400 miles @ 11½ kts. Draft, 4'3'' for'd, 7'6'' aft.
Protection	Blast screen and wooden shelter on bridge.
Propulsion	As in LCT (2), (3).

UK

LCA (HR) LANDING CRAFT, ASSAULT (HEDGEROW)

Operational use — To explode minefields in the path of landing forces.

Description — A British conversion of the LCA, Landing Craft, Assault (over-all photo). As shown by the detail photographs, very few alterations beyond strengthening some hull frames were found necessary. All statistics are identical to the LCA.

Armament — 4 rows of 6 mortars.

UK

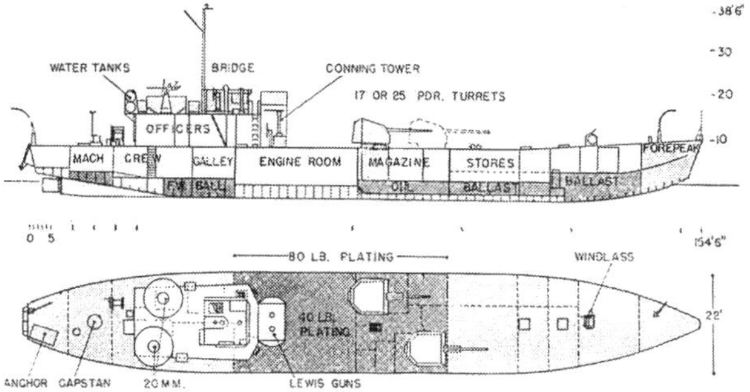

LANDING CRAFT, GUN (MEDIUM) **LCG (M)**

Operational use	To engage and neutralize or destroy coastal installations.
Description	Medium-sized version of LCG (L), designed and built for the purpose rather than converted, as in other craft.
Dimensions	Length, 154'6'' o.a. Beam 22'4''. Draft, 5'6'' (Operating)
Displacement	270 tons light 6'3½'' (mean) loaded
Speed	13½ kts. (max.).
Armament	Two 17-pdr. or 25-pdr. Army guns; two 20-mm. and two smaller M.G.
Armor	2'' NC on sides and control station; 1'' NC on deck over engines and mag.; 15-lb. plating to gun positions and conning tower.
Propulsion	2 Paxman 460-500 hp. Diesels.
Fuel	14,230 gals.
Crew	3 officers, 27 men.

UK

LCG(L)(3)

Photo below shows LCG (L) (3). For further details see following page.

LCG(L)(3) LANDING CRAFT, GUN (LARGE) (MARK 3)

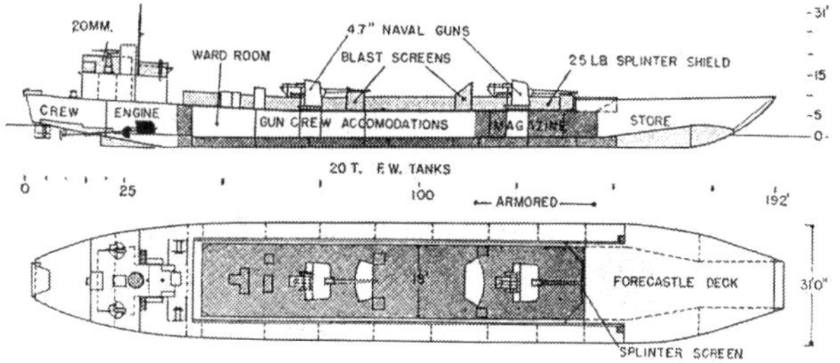

Operational use	To provide direct fire against beach positions and surface attack for first-echelon landing waves. Craft is beached to become a stationary gun platform.
Description	British conversion from LCT (3) hulls. Ships have been made watertight; had false deck built over hold. Drawing illustrates rearrangement for magazine space and additional accommodations.
Endurance	2,700 miles at 10½ kts. Speed, 10½ kts. (max.).
Dimensions	Length, 192'0" o.a. Beam, 31'0".
Displacement	500 tons (loaded). Draft, 3'6" for'd, 6'0" aft.
Armament	Two 4".7 QF Naval guns; two 20-mm. or two 2-pdr. pompoms.
Armor	25-lb. plating to side of gun deck, blast screens, sides of magazines. (See drawing.)
Crew	2 officers, 13 men; gun crew, 1 officer, 2 NCO's, 20 enlisted marines, with full living accommodations.
Propulsion	2 Paxman 500 hp. Diesels, twin screw.
Fuel	24 tons Diesel fuel.

UK

LCG (L) (4) LANDING CRAFT, GUN (LARGE) (MARK 4)

It is similar in appearance to the Mark 3; converted from LCT (4) hulls.

UK

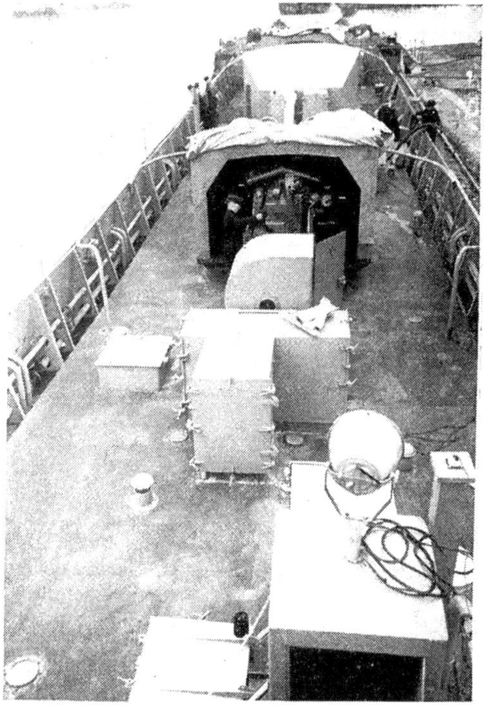

MOTOR GUNBOAT **PGM**

Operational use	To furnish close fire support to a landing operation. Also to intercept and destroy inter-island barge traffic.
Description	Converted from SC hulls with few alterations in design.
Dimensions	110′ o.a. x 18′ beam.
Displacement	108 tons. Draft, 5′5″ (mean). 6′9″ (loaded).
Armament	One 3″/23 or one 40-mm. for'd. One 40-mm. aft. Four .50 cal. twins. One 60-mm. Army mortar. One multibarrel (6) bazooka (60 rockets carried). Two FM smoke generators.
Armor	¼″ splinter protection to gunshields, control station.
Endurance	690 miles @ 19 kts. (max.), 2300 @ 12 kts., 2,890 @ 10 kts.
Propulsion	Two 1200 hp., G. M. Diesels, twin screws and rudders.

★ ★ ★

A similar conversion is authorized for the 173′ PC.

Armament	One 3″/50; one 40 mm. twin; six 20 mm.; one 60 mm. mortar; one 6 barreled bazooka.
Armor	½″ plating to control, hull sides at engine bulkhead in way of engine room, ¼″ splinter protection to all guns.

US

LCC (1)

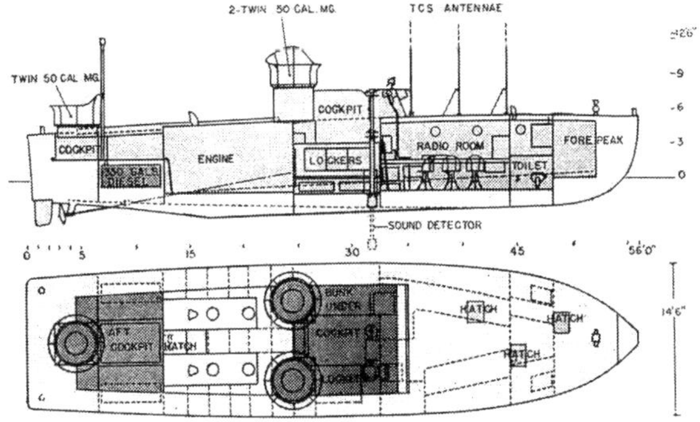

LANDING CRAFT, CONTROL (MARK 1) **LCC (1)**

Operational use	As lead-in navigational craft for landing boats; to mark line of departure; for traffic control; for preliminary hydrographic surveys.	
Description	Similar in appearance to the LCS (S), this steel-hull craft is distinguished by its collapsible antennae and radar. Usually carried on the decks of APA's and AKA's and lowered by booms. Not intended to beach.	
Endurance	500 miles @ 10 kts; 240 miles @ max. speed.	Speed, 13½ kts. (max.).
Dimensions	Length, 56'0" o.a. Beam, 13'7".	
Displacement	30 tons. Draft, 3'11½" (max.).	Capacity, crew only.
Armament	3 twin .50-cal. M.G. on ring mounts; smoke pots.	
Armor	¼" STS on bridge and gun cockpits.	
Special Equipment	Gyro compass.	Odograph (small DRT).
	SO radar w/VPR scope.	QBG underwater sound equipment.
	NK-2 echo sound receiver.	ZB/RU radio directional equipment.
	NJ-8 echo sound receiver.	2-TCS radio transmitter/receivers.
	Gyro repeater.	2-SCR 610 (FM radio transmitter/receivers.)
Crew	14	
Propulsion	2 Diesels, 225 hp. each.	Fuel, 550 gals. fuel oil.

US

LANDING CRAFT, CONTROL (MARK 2) **LCC (2)**

Differs from LCC (1) in draft (3'11" max.), armament (2 twin .50-cal.) and SO radar (has NPM microfilm chart projector).

US

LCC LANDING CRAFT, CONTROL

Operational use	To serve as navigational lead boat for groups of landing craft.
Description	Converted from British FAIRMILE "B" Motor Launches.
Dimensions	Length, 112′ Beam, 18′6″
Displacement	75.5 tons. Draft, 5′ aft (max.)
Speed	20 knots, 17 knots cruising speed.
Armament	One 3 pdr. Mk. V. Two 20 mm. (Mark 3, 4). One Y-gun, ?-DC carried. CSA smoke generator. Two Vickers Mk. V MG, 1-twin MG.
Propulsion	2 gas engines @ 1,120 hp.
Fuel	2,330 gals.
Equipment	QH2 navigational aid. DF loop control. Echo sounding recorder. Asdic recorder. Pitometer or Chernikeef log recorder. Standard compass. 9 or 20 mile measuring reel.

UK

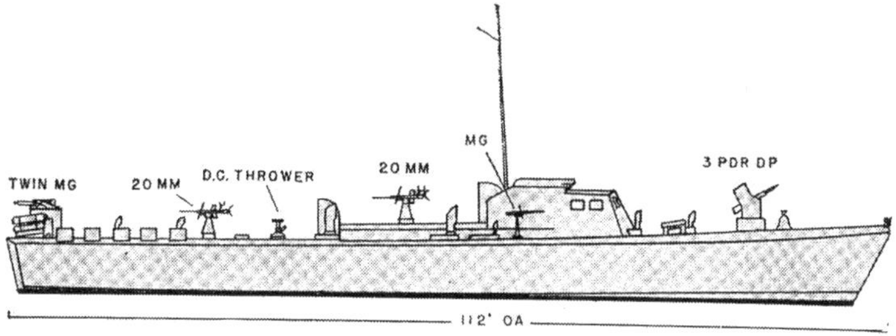

LCN LANDING CRAFT, NAVIGATIONAL

British conversion of the LCP (L) to act as navigational guide to groups of landing craft. Similar in appearance, it differs from the LCP (L) only in additional navigational equipment.

UK

LCH LANDING CRAFT, HEADQUARTERS

A British conversion of LCI (L) designed to act as communications headquarters for groups of landing craft. Alterations to the mast are the only external appearance changes.

UK

LBV(1)	Landing Barge, Vehicle (Mark 1)
LBV(2)	(Mark 2)
*LCV	Landing Craft, Vehicle
*LCVP	Landing Craft, Vehicle, Personnel
LCM(1)	Landing Craft, Mechanized (Mark 1)
LCM(2)	(Mark 2)
*LCM(3)	(Mark 3)
LCM(4)	(Mark 4)
LCM(5)	(Mark 5)
*LCM(6)	(Mark 6)
LCM(7)	(Mark 7)
*LCT(1)	Landing Craft, Tank (Mark 1)
*LCT(2)	(Mark 2)
*LCT(3)	(Mark 3)
*LCT(4)	(Mark 4)
*LCT(5)	(Mark 5)
*LCT(6)	(Mark 6)
LCT(8)	(Mark 8)
LCT(9)	(Mark 9)
Z-Lighter	

*Major operational types.

ONI 226 : CONFIDENTIAL
Publication and Distribution Branch—Division of Naval Intelligence

LANDING BARGE, EMERGENCY REPAIR LBE

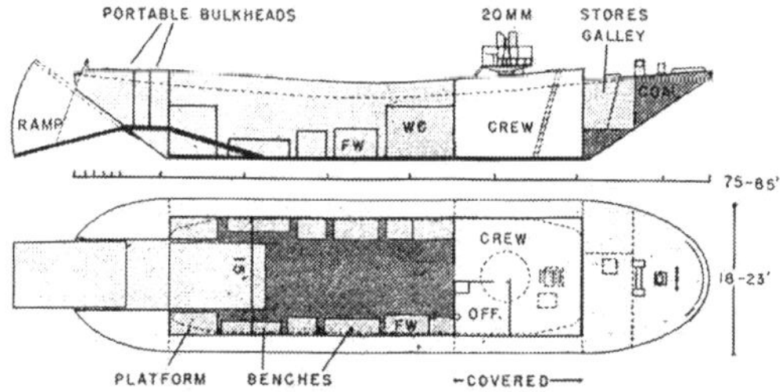

Operational use	Small floating repair units for the maintenance of minor landing craft in a beach area. Mobile repair facilities are later debarked onto the beach.
Description	Converted steel commercial dumb barges of varying dimensions, equipped with a stern ramp. Similar to the LBO, LBF, and LBV.
Capacity	3-ton workshop lorry (No. 4 Mk. 3).
Dimensions	75'–85' x 18'–23'.
Draft	4'–4'6'' mean (loaded).
Lorry Equipment	6'' lathe; hand and electric drilling machines; portable forge and anvil; bench and 2 vises; 7½ kw. generator and charging board.
Barge Equipment	2 benches, each with 6'' vise; portable electric welding set; oxy-acetylene welding-cutting set including 6 oxygen, 4 acetylene bottles. Stowage for spare batteries for craft. Coal stowage. 4 tons fresh water carried.
Endurance	360 miles @ 5 kts. Speed, 6 kts. (max.) on powered units.
Armament	One 20 mm.
Armor	10-lb. steel plating over cabin.
Crew	2 officers, 16–22 men with accommodations; 6–10 additional men in hammocks.
Propulsion	Being fitted with 2 Chrysler RM gasoline engines, single rudder.

UK

LBO LANDING BARGE, OILER, LBW LANDING BARGE, WATER

Operational use	Small barges equipped to refuel or supply minor landing craft in a beaching area.
Description	Converted from Thames commercial steel barges of various dimensions. Absence of ramp distinguishes them from other conversions.
Capacity	40 tons in tank; some are divided into 2 sections for different types of fuel. Fitted with two 5" hand pumps with capacity of 1,800 gals./hr. each. Tanks are filled by 2" Admiralty screwed hose. Fire extinguishers are controlled from the upper deck.
Speed	6 kts. (max.) on powered types. Endurance 360 miles @ 5 kts.
Draft	4'6"–6' mean (loaded). Armament 1 twin .303 Lewis gun.
Armor	2½" plastic over fuel and gasoline tanks and to steering shelter; 3" concrete sides to hold; 5-lb. plating over engine; 10-lb. WT bulkheads in ramp entrance and forward of engine room.
Propulsion	2 Chrysler RM gas engines and single rudder being fitted.

UK

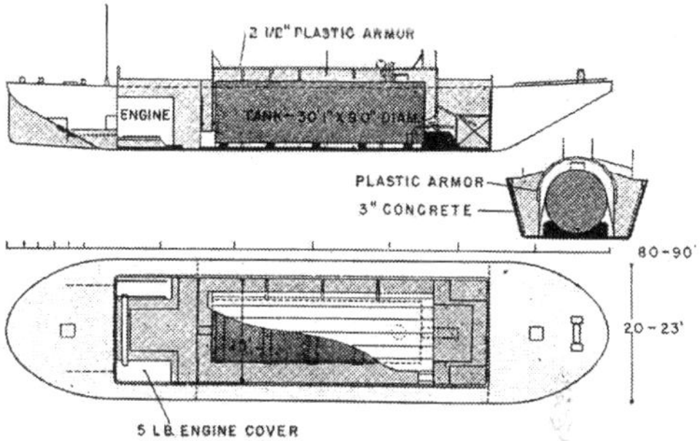

LANDING BARGE, KITCHEN LBK
(British Staff Requirements)

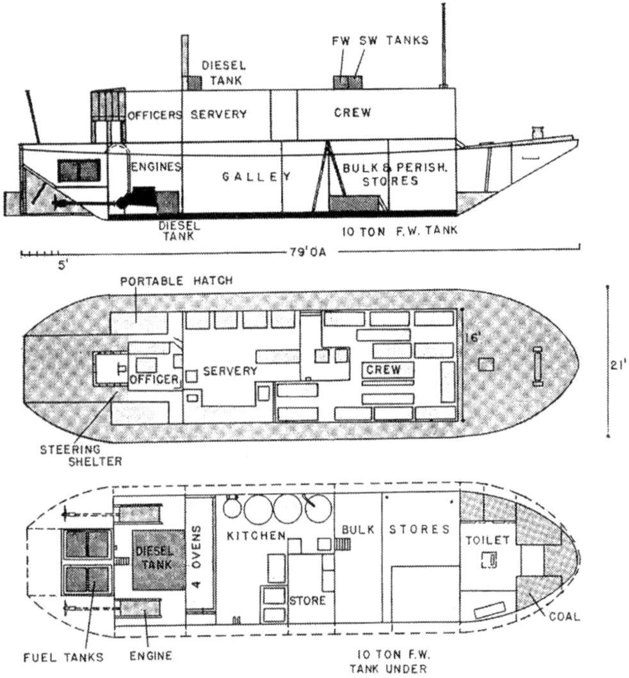

Operational use	To serve as floating kitchen to groups of landing craft.
Description	Converted from steel swim barges, similar to the LBE—LBO.
Capacity	Storage and issuing space for enough provisions to feed 900 men for one week.
Dimensions	79′ o. a. x 21′ x 3′6″ draft (estimated.)
Armament	None.
Armor	2½″ plastic to Diesel tanks, engines, steering shelter.
Crew	1 officer, 24 men.
Endurance	300 miles at 5 knots (estimated).
Propulsion	2 Chrysler engines.
Fuel	600 gallons gasoline.

LCE LANDING CRAFT, EMERGENCY REPAIR

Operational use	Small salvage and repair craft equipped to provide temporary repairs to disabled minor landing craft in a beach area.
Description	British conversions of the LCV, these craft are carried on deck or on davits with the same sling spacing as the LCA.
Endurance	80 mi. @ 9 kts. (max. speed). Draft 3′
Equipment	1 Coston Line Gun One 90 gals./min. salvage pump with one 10′ No. 2 hose, two 30′ No. 2 hose. One 80-lb. anchor with 40 fathoms 2¼″ rope. 50′ of 2 x 2″, 6 x ½″, 4 x ½″ timber. 24 wedges. 10 sheets of waterproof plywood. ¼ bolt of heavy canvas. 12 sheets of 4′ x 1′8″ x 16 oz. copper sheeting. Nails, screws, lalton, spare engine parts, tools, etc.; Fire-fighting equipment—No. 2 Branch 2 spray nozzles. 4-gal. knapsack foam tank. 20 gals. refills. Explosives box.
Armor	10-lb. DIHT plating to engine sides, bulkheads, hull sides, linatex fuel tanks.
Crew	1 officer, 3 men; 2 motor mechanics, 2 shipwrights, 1 wireman, 1 stoker.
Weight	23,500 lb. (loaded) (max.) All other statistics are identical to the LCV.

UK

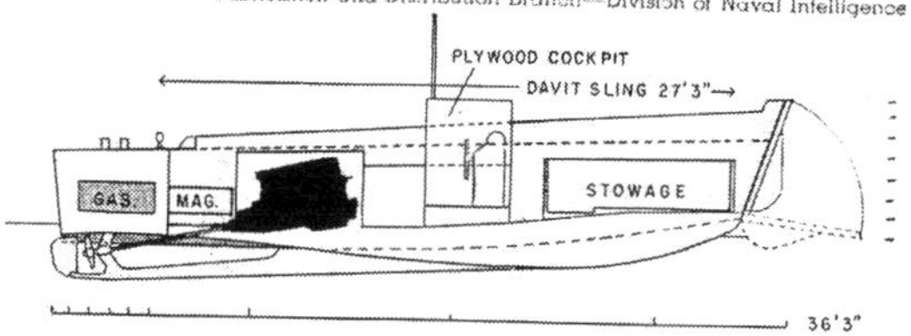

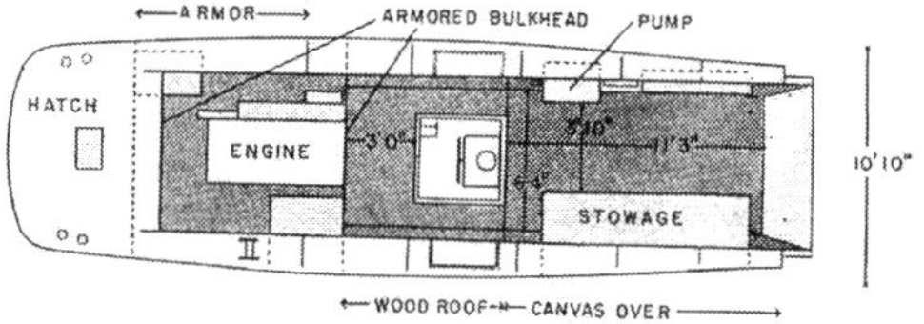

AMPHIBIOUS VEHICLES

LVT (1)	Landing Vehicle, Tracked	(Mark 1)	
*LVT (2)		(Mark 2)	
*LVT (3)		(Mark 3)	
*LVT (4)		(Mark 4)	
*LVT(A) (1)		(Armored)	(Mark 1)
*LVT(A) (2)		(Armored)	(Mark 2)
LVT(A) (3)		(Armored)	(Mark 3)
*LVT(A) (4)		(Armored)	(Mark 4)
*DUKW	2½-ton, 6 x 6, Amphibian Truck		
*JEEP	¼-Ton, 4 x 4, Amphibian Truck		
TERRAPIN 1			
TERRAPIN 2			

*Major operational types

LANDING VEHICLE, TRACKED (MARK 1) — LVT (1)

Operational use	Personnel and cargo carrier designed for landing operations over coral reefs and other obstacles.
Description	Steel constructed watertight hull propelled on land and in water by cleated tracks. Transported on deck or in the hold of LST and LSD. LVT (1) is now obsolete.
Capacity	4,500 lb. (normal) or 20 fully equipped men; cargo space, 8'6'' long x 7'9'' wide x 3'11'' deep.
Endurance	75 miles (land); 50 miles (water).
Speed	15 m.p.h. (land); 4 knots (water) (loaded).
Armor	None.
Dimensions	21'6'' o.a. x 9'10'' x 8'2'' height o.a.
Weight	16,900 lbs. (unloaded).
Armament	Track rails for one .30 cal. and one .50 cal. M.G.
Crew	3.
Propulsion	Hercules WXLC three 6-cylinder, 146 hp. @ 2,400 r.p.m. gasoline engine; 80 gals. fuel carried.

US

LVT (2)

ONI 226 · CONFIDENTIAL

Publication and Distribution Branch—Division of Naval Intelligence

LANDING VEHICLE, TRACKED (MARK 2) LVT (2)

An improved version of LVT (1), this type is now standard for the United Nations. Soon to be superseded in production by LVT (4).

Capacity	6,500 lbs. (normal), or 24 fully equipped men.
Endurance	150 miles (land), 75 miles (water).
Speed	25 m.p.h. (land); 5.4 kts (water).
Armor	Portable plating shown by diagram.
Dimensions	26'1" o.a. x 10'8" x 8'1" height o.a.
Weight	25,200 lbs. (unloaded).
Armament	Track rails for 1– .30 and 1–.50 cal. M.G.
Crew	3.
Propulsion	Continental radial air-cooled 7-cylinder 200 hp. @ 1,800 r.p.m. gasoline engine.
Fuel	110 gals. gasoline carried — consumption — 13 gals. an hour.

US

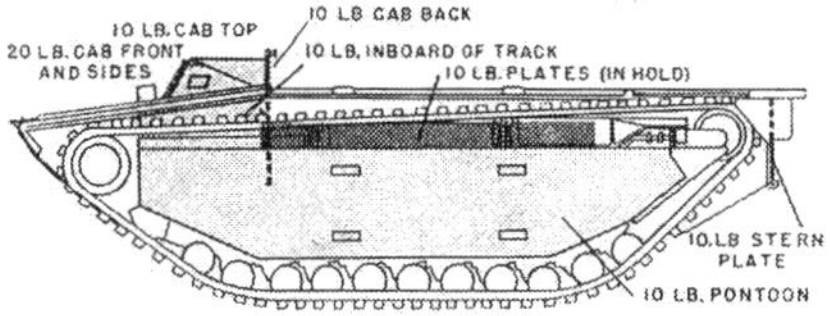

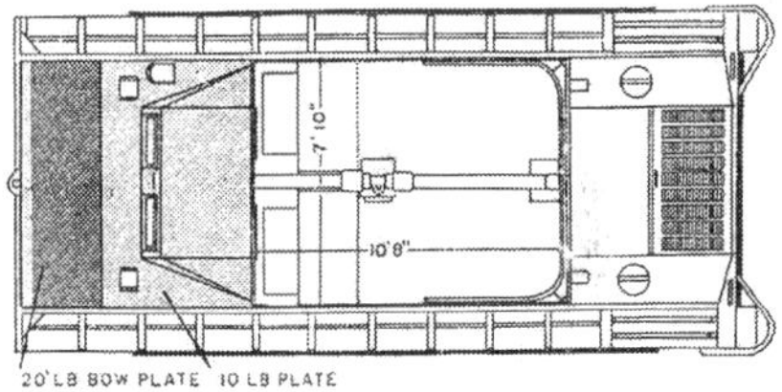

LVT (3) LANDING VEHICLE, TRACKED (MARK 3)

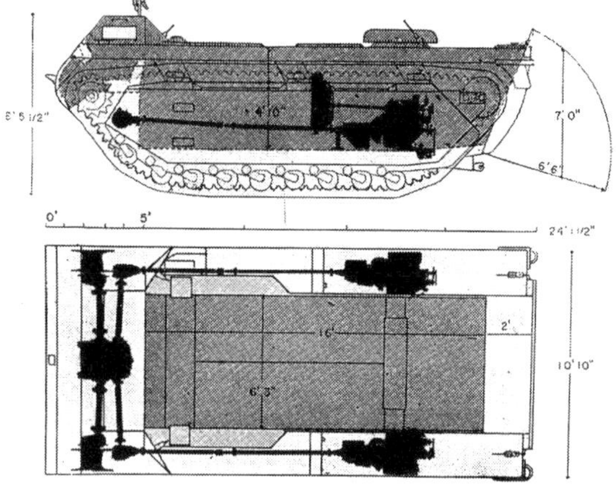

Capacity	8,000 lbs. or 24 fully equipped men. Ramp is capable of 5,000 lbs. weight.
Endurance	150 miles (land), 75 miles (water).
Speed (est.)	5.2 kts. (water), 25 m.p.h. (land).
Armor	Portable.
Dimensions	24'1½" o.a. x 10'10" x 8'5½" height o.a.
Weight	28,000 lbs. (unloaded).
Armament	One .30 cal. and one .50 cal. M.G.
Crew	3.
Propulsion	Two Cadillac V-8 water-cooled 220 hp. @ 3,400 r.p.m.—gasoline engines.
Fuel	150 gallons gasolire.

US

LVT (4)

LANDING VEHICLE, TRACKED (MARK 4) LVT (4)

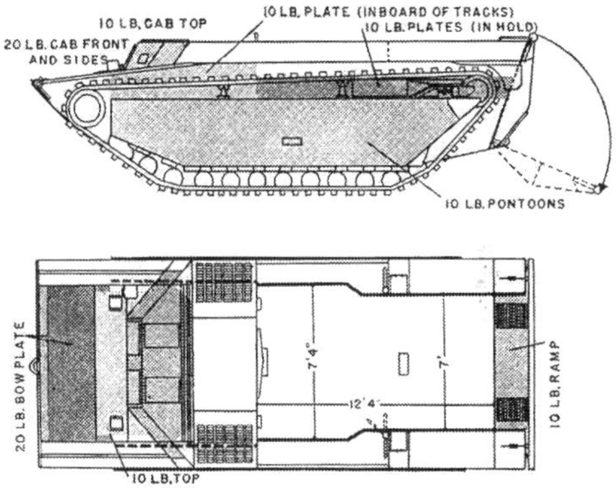

An LVT(2) with a stern ramp.

Capacity 6,500 lbs. (max.).

Speed Same as LTV (2).

Armor Removable armor kits are illustrated on diagram.

Weight 23,350 lbs. (unloaded).

Dimensions 26'1'' o.a. x 10'8'' x 8'2½'' height o.a.

Propulsion Same as LVT (2).

US

LVT (A) (1) LANDING VEHICLE, TRACKED (ARMORED) (MARK 1)

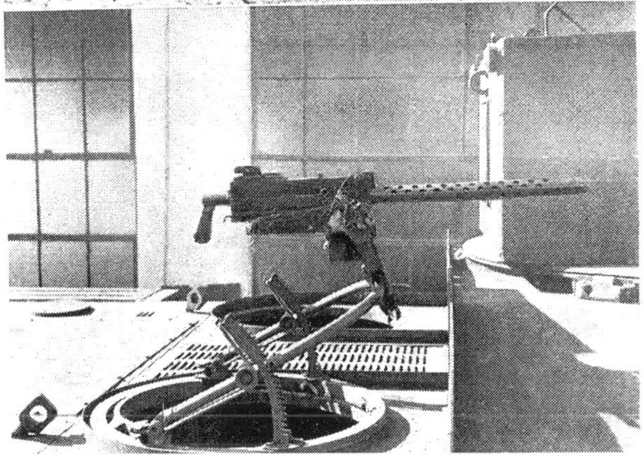

LANDING VEHICLE, TRACKED (ARMORED) (MARK 1) LVT(A)(1)

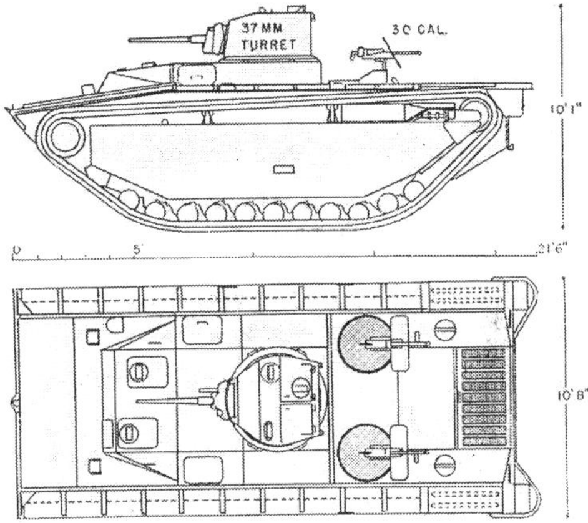

A lightly-armored amphibious combat vehicle, similar in characteristics to the LVT(2).

Capacity	Maximum of 1,000 lb.
Dimensions	26'1" o.a. x 10'8" x 10'1" height o.a.
Weight	32,800 lbs.
Armament	Turret, one 37 mm. M-6 gun and One .30 cal. coaxially mounted; Two .30 cal. M.G. on ring mounts.
Armor	½" vertical cab sides, turret, and under bow; ¼" elsewhere.
Crew	6.

All other data is identical to LVT(2).

US

LVT (A)(2) LANDING VEHICLE, TRACKED (ARMORED) (MARK 2)

An armored LVT(2) capable of carrying 6,500 lbs. of cargo. Except for weight (27,600 lbs.), all statistics are the same as LVT(2). Amor and crew are identical to LVT(A)(1).

LVT(A)(3) was designed as an LVT(4) with built-in armor, but will not be produced.

US

LANDING VEHICLE, TRACKED (ARMORED) (MARK 4) LVT(A)(4)

An LVT(A)(1) with an Army M-8 75 mm. Howitzer turret replacing the 37 mm. Only one .50 cal. is carried as additional armament. Excepting the statistics shown below, all data is identical with the (A)(1).

Weight	38,000 lbs.	Capacity 2,000 lbs. ammunition and gear.
Speed	25 mph. (land) 5.2 kts. (water)	Crew 5.
Armor	1'' turret, ½'' vertical cab sides, ¼'' otherwise.	

US

▲ Tandem arrangement permits ferrying bulk cargoes, tanks, aircraft, etc.

▼ British LSI(L) unloading by davits.

2½-TON, 6 x 6, AMPHIBIAN TRUCK **DUKW**

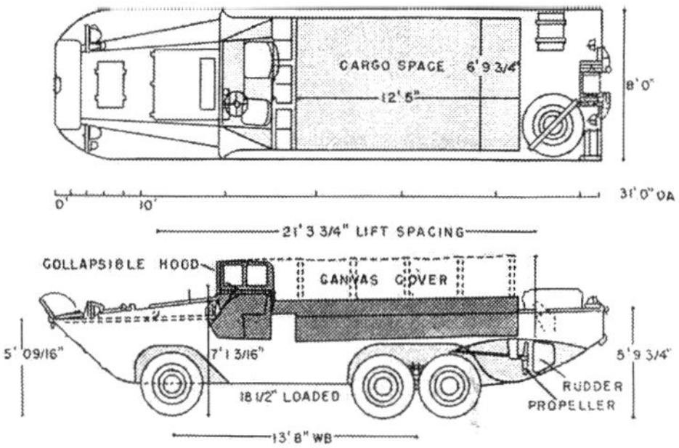

Operational use	Amphibious Army truck for ship-to-shore transport, capable of operating in a moderate sea and surf.
Description	6-wheeled truck with a boat hull, propeller in tunnel, and small rudder. Hold may be covered by portable canvas and litters; cockpit has collapsible hood. Sometimes carried at davits, generally transported on the decks of ships.
Capacity	25 troops and equipment or 12 loaded litters, or 5,000 lbs. of cargo, all limited by freeboard.
Endurance	400 miles at 35 miles per hour on land.
Speed	5.5 kts. in water; 50 m.p.h. (max.) on land.
Displacement	Light, 13,000 lbs. Loaded, 18,600 lbs.
Propulsion	6-cylinder GM engine; 6-wheel drive on land; propeller in water; steered in water by front wheels and small rudder.
Crew	One.

US

JEEP ¼-TON, 4 x 4, AMPHIBIAN TRUCK

An amphibian model of the Army "Jeep," capable of operating in moderate seas and surf.

Dimensions	15'7" o. a. x 5'4".	Capacity, 800 lbs. (incl. 1 crew).
Weight	3,700 lbs. (light).	Endurance, 300 miles (land).
Speed	50 m.p.h. (land); 4.7 kts. (water).	

These all-purpose vehicles are often equipped with various types of armament, including rockets.

US

TERRAPIN (Mark 1)

Operational use	To operate between ships and shore dumps during the establishment of a beachead; also to act as a transport for wounded from shore to ship.
Description	British 8 x 8 amphibious truck capable of operating in moderate surf and all beach conditions.
Capacity	5 tons in calm water, 4 tons in medium water, 50 men with full equipment.
Weight	15,450 lbs. (unloaded).
Speed	15 m.p.h. (max. land).
Dimensions	Length, 23' o. a. Beam, 8'9". Height, 8'9".
Propulsion	2 Ford V-8 engines.

UK

TERRAPIN (Mark 2)

(British staff requirements)

Description	Designed to overcome disadvantages revealed in trials of earlier models.
Capacity	11,200 lbs. (13,440 desired).
Weight	Not to exceed 22,400 lbs.
Speed	20 m.p.h. (land), 6-7 knots (water).
Dimensions	Length, 31'. Beam, 8'6". Height, 8'10"
Endurance	125 miles (land), 4 hours (water).

UK

*LSM	Landing Ship, Medium
LST	Landing Ship, Tank
LST(1)	Landing Ship, Tank—(Class 1)
*LST	
LST(3)	(Class 3)
LST(4)	(Class 4)
*LSD	Landing Ship, Dock
ARL	Repair Ship, Landing Craft
LSE	Landing Ship, Emergency Repair
LSV	Landing Ship, Vehicle
LSS	Landing Ship (Stern-chute)
LSC	Landing Ship, Carrier (Derrick-Hoisting)
LSG	Landing Ship, Gantry
*LSI(H)	Landing Ship, Infantry (Hand-Hoisting)
LSH	Landing Ship, Headquarters
LSI(S)	(Small)
LSI(M)	(Medium)
LSI(L)	(Large)
LSP	Landing Ship, Personnel
LSB	Landing Ship, Bombardment
LSF	Landing Ship, Fighter Direction
*AGC	Amphibious Force Flagship
*APD	High Speed Transport (Destroyer)
*APC	Coastal Transport

*Major operational types

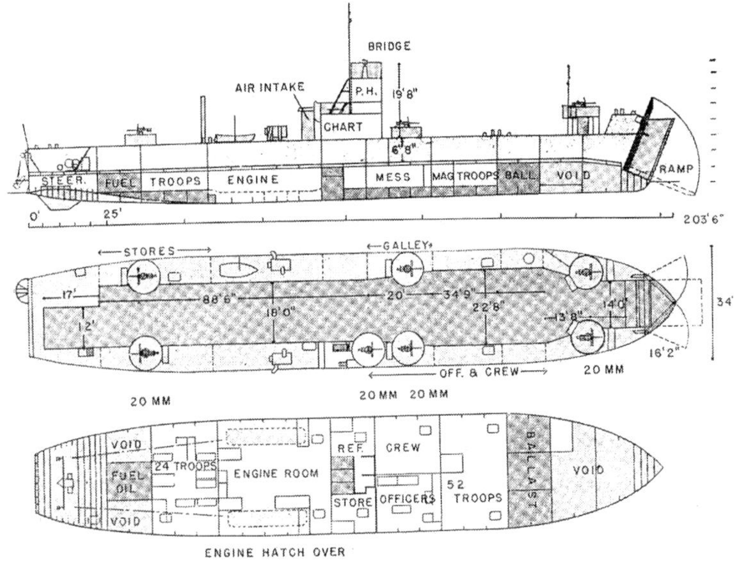

Operational use — An oceangoing tank landing ship designed to operate with LCI (L)'s.

Description — A new design, derived from a combination of the LST and LCT (6), and formerly designated LCT (7).

Capacity — 5 medium or 3 heavy tanks, or 6 LVT or 9 DUKW's. (150 max. load to maintain landing draft). Troops—2 officers, 52 men.

Endurance — 3,500 mile radius @ 12 knots (estimated).

Speed — 13.3 kts. @ 720 r.p.m. (designed).

Displacement — Seagoing (full load), 900 tons.
Draft, for'd, 4'5'' aft 7'11$\frac{1}{2}$''
Landing (full load), 741 tons.
Draft, for'd, 3'5'' aft 6'11$\frac{1}{4}$''
Seagoing (light) 513 tons.

Tons per inch immersion at landing disp'l.—14.10; at oceangoing disp'l. 14.3

Dimensions — Length, 203'6'' o. a. Beam, 34'0''. Armament 6–20 mm. mounts

Armor — 10-lb. STS, splinter shield for 20 mm. guns and sides of pilothouse.

Crew — 4 officers, 48 men.

Propulsion — Fairbanks-Morse Diesel or General Motors Cleveland Diesel direct drive 2,800 hp., @ 720 r.p.m.

US

LSM
LANDING SHIP, MEDIUM

LSM

LST LANDING SHIP, TANK

BACHAQUERO Class
BACHAQUERO, MISOA.
TASAJERA.

British conversions of MARACAIBO oilers, designed to transport and land tanks directly on steep beaches.

Dimensions	BACHAQUERO Class, 382'6" o.a. x 64'0" x 10' mean draft. TASAJERA, 365'0" o.a. x 60'2" x 10' mean draft.
Capacity	18 Churchills or twenty-two 25-ton tanks or 33 trucks; 2 LCM or smaller on deck.
Speed	9–10 knots (max.).

See tables for further information on both types.

Photographs are of BACHAQUERO units.

UK

LANDING SHIP, TANK (Class 1) LST (1)

Operational use	A British-designed, oceangoing, tank-carrying ship, capable of landing medium tanks over a bow ramp (and extension) directly on a beach steeper than 1/37.
Description	Designated as the "Boxer Class"—consisting of 3 ships—Boxer, Bruiser, Thruster. Troops and crew are accommodated on the first deck, tanks on the second. Notice stack is off-center to allow for continuous tank deck in hold. A 40-ton crane is carried to handle minor landing craft.
Capacity	Thirteen 40-ton or twenty 25-ton tanks and 36 loaded 3-ton lorries (150 tons total). Troops, 13 officers, 180 men.
Dimensions	390'0" b.p. x 49'0" max. beam.
Endurance	8,000 miles @ 14 kts. (estimated). Beaching fuel load is 300 tons.
Speed	17 knots (max.).
Displacement	Loaded, 5,410 tons. Landing, 3,616 tons.
Draft	5' for'd, 13' aft. (landing).
Armament	Twelve 20 mm.; two 4" mortars.
Crew	25 officers, 137 men.
Propulsion	T. S. Turbine, 7,000 h.p.
Fuel	2,100 tons (max.).

UK

LST

Photos show standard 2–davit design with old armament, since reinforced as shown by the diagrams. These illustrate the ultimate armament for both the 6–davit and 2–davit designs.

LANDING SHIP, TANK—LST

Opera-tional use Ocean-going ship designed to land waterproofed tanks or vehicles over a low ramp on a 1/50 beach slope.

Description An American design now the United Nations standard. Elevator and hatch service main deck where miscellaneous vehicles and cargo are stowed. Starting with LST-513 and excepting LST-531, main deck ramp is substituted for elevator. For inaccessible landings, sectional pontoons are used, or earth causeways built. LCT(5) or (6) may be carried in sections or as a unit on main deck. UK Designation—LST(2).

Capacity Ocean-going load is limited to 2,100 tons on LST(1) class and 1,900 tons on LST-542 class. Of this maximum, main deck load is 350 tons. Main deck is designed for concentrated load of trucks having 10-ton gross weight. When this load is exceeded, as in case of LVT's, vehicles should be loaded over short span side beams with planking to distribute load.

Tank deck load is designed for concentrated load of heaviest tanks. The limiting height from deck to underside of lights in this space is 11'3''. Volume is 92,765 cubic feet. 1,060 tons of Diesel oil can also be carried.

Endurance 6,000 mile radius at 9 knots. Speed—10.8 knots (maximum).

Displace-ment

Sea-going (full load of 2,100 tons) 4,080 tons.	Draft, 8' fwd. 14'4" aft.
Landing (full load of 500 tons) 2,160 tons.	Draft, 3'1" fwd. 9'6" aft.
Light, 1,490 tons.	Draft, 1'6" fwd. 7'5" aft.

Tons per inch immersion Landing Displacement 33.3
Tons per inch immersion Ocean-going 33.5
LST-542 class has a light displacement of 1,623 tons. The increased displacement is a result of the additions of authorized armament, a distilling plant (4,000 gallons per day), and other miscellaneous weight increases. Payload to maintain designed landing draft is decreased accordingly.

Dimensions Length, 328'0" o.a. Beam, 50'0"

Armament U. S. 1-3"/50 cal. D. P. (omitted when 40's are available) 1-40 mm. (7 authorized). 6-20 mm. (12 authorized).
BRITISH: 1-12 pdr. 6-20 mm 4-FAM

Armor 15-lb. STS splinter protection.

Crew (2 davit)—7 officers, 204 men. Troops—16 officers, 147 men.
(6 davit)—9 officers, 220 men. Troops—14 officers, 131 men.

Propulsion Two 900-hp. Diesels; twin screw.

US

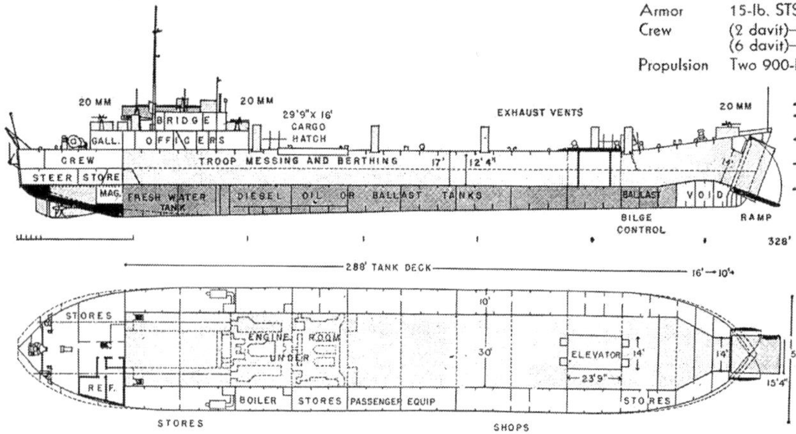

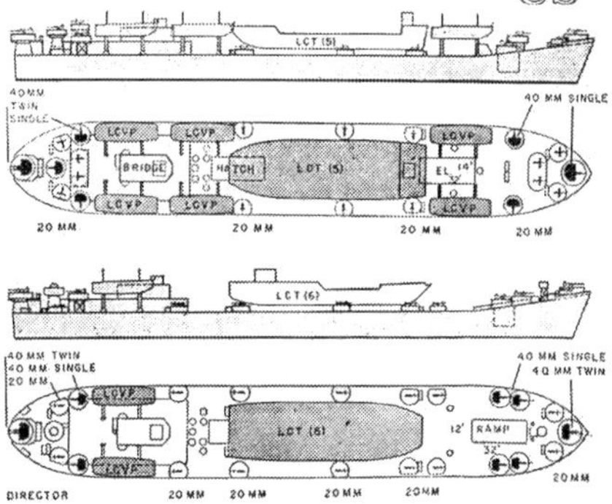

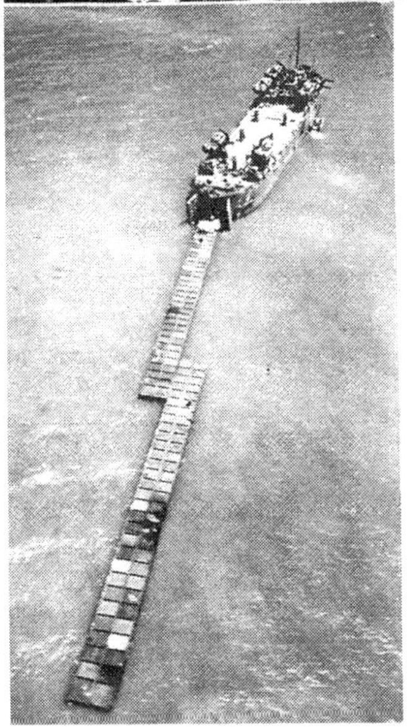

◄ ▲Photos show sectional ramp extension carried on deck and in use. ▲ 6-davit LST ▼LCT and Army MTL's carried on deck.

LANDING SHIP, TANK (CLASS 3) LST(3)
(British Staff Requirements)

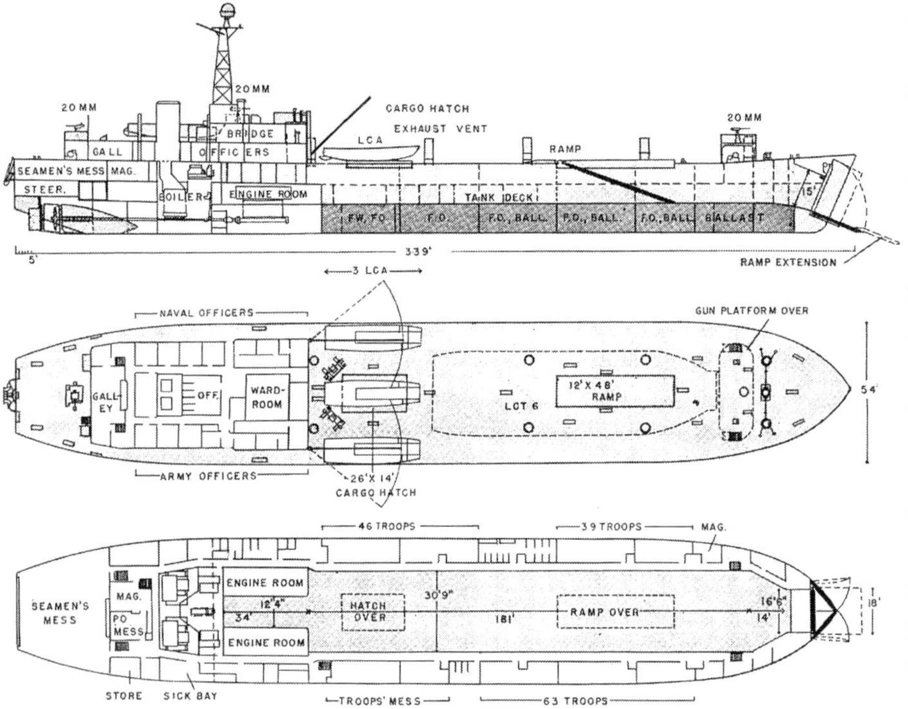

Operational use
: Oceangoing tank-landing ship capable of landing tanks and vehicles directly on the beach.

Description
: An oceangoing British adaptation of the American LST design, incorporating steam propulsion. Building in the Far East.

Capacity
: 2 LCA at davits, 3 LCA in chocks, 1 LCT (6) at expense of vehicles on upper deck, 14 trucks on upper deck, 27 25-ton tanks or 15 40-ton tanks on tank deck; troops – 18 officers, 150 enlisted.

Dimensions
: Length, 345'10'' o.a. Beam, 54'.

Displacement
: Loaded, 4,820 tons.
: Beaching, 3,065. Draft, 4'7'' for'd, 11'6'' aft.

Armament
: 4 twin 20 mm. guns and 2–20 mm. single.

Armor
: 15 lbs. DIHT around bridge, pilothouse and gun positions.

Endurance
: 8,000 miles @ 11 kts. (loaded).
: 1,000 miles @ 10 knots (Landing condition.)

Speed
: 13.5 kts. (at beaching displacement).

Crew
: 14 officers, 90 enlisted.

Propulsion
: Two 2,750 hp. reciprocating steam engines.

UK

Photos show BELLE GROVE in a light condition. 8-16-43.

LANDING SHIP, DOCK **LSD**

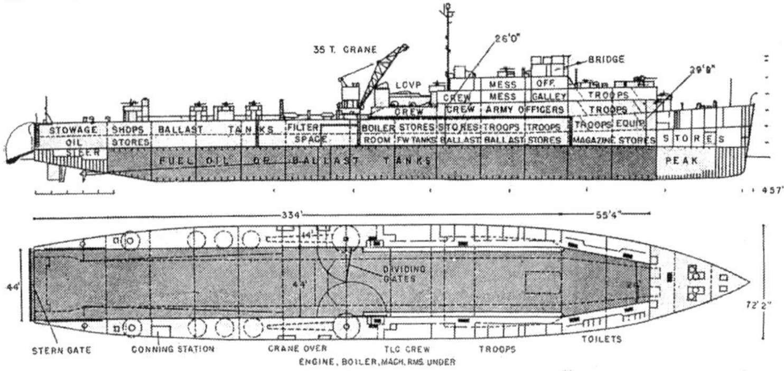

Operational use To transport loaded landing craft to the landing area, where the hold is flooded and the craft move out under their own power.

Description An adaptation of the floating dry-dock principle. Landing craft up to the size of LCI (L)'s may be stowed in the well; 2 35-ton cranes service the craft, handle vehicles and cargo. British Eastway Class.

Capacity 3 LCT (5) (6) each with 5 medium tanks or 2 LCT (3) (4) each with 12 medium tanks or 14 LCM (3) each with 1 medium tank or 1,500 long tons of cargo or 41 LVT's or 47 DUKW's. Troops, 22 officers, 218 men.

Endurance 8,000 miles @ 15 knots. Design speed is 17 knots maximum.

Displacement

		Draft	Draft
Light draft	4,032	8'2½" for'd	10'1½" aft
Light service	5,850	10'8"	13'8"
Seagoing (loaded)	7,930	15'5½"	16'2"
Ballasted (no cargo)	14,078	30'9½"	29'9½"

Well draft 8'1" for'd, 9'11" aft

Tons per inch immersion, light condition 43
loaded condition 55

Dimensions Length 457'9" o.a.; Beam 72'.

Armament

U.S.	British
1–5"/38 cal. DP	1–3"/50 cal. DP
2–40 mm. twin	4–2 pdr. pompoms
2–40 mm. quad.	16–20 mm. singles
16–20 mm. singles	

Armor Splinter protection for gun emplacements and pilot-house sides.

Crew 17 officers, 237 men; LC 6 officers, 30 men.

Propulsion LSD 1–8, steam reciprocating; LSD 9–15, steam turbines; twin screws.

US

LSD TEMPORARY SUPERDECKS—FOR CARRYING LVTs AND DUKWs

Capacity with two temporary decks (with ramps)

	Tank Deck	2d Deck	3d Deck	Total
LVT	41	33	18	92
DUKW	47	40	21	108

Capacity with one temporary deck (no ramp)

	41	0	23	64
	47	0	27	74

Two temporary decks which are subassembled will be installed in LSD 13–27. The top temporary deck is supported by the side weather decks. The intermediate or 2d deck is supported on stanchions and can be removed independently of the top or 3d temporary deck to permit loading of LCMs. Ramps are provided at the after end which can be used to discharge LVTs and DUKMs as long as the intermediate deck is installed. Otherwise, top deck is extended aft without ramps and vehicles must be unloaded by one or both of ship's 35-ton cranes.

A Bridge crane is to be installed in LSD 13–27 which will facilitate assembling of temporary deck in enclosed portion of LSD well.

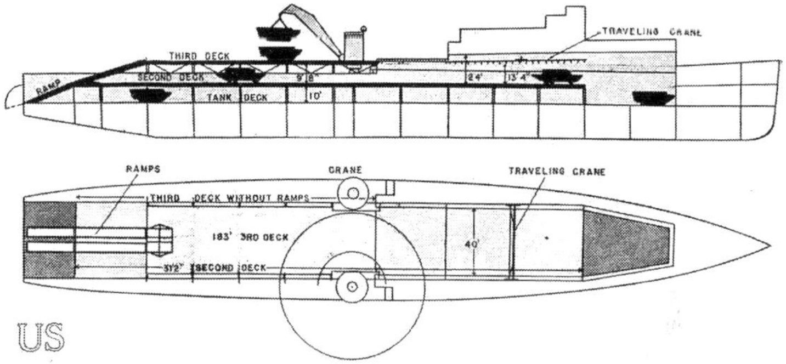

Capacity:

	LVT	DUKW	TROOPS
AP 106–107	None	44	800
*AP 108–109	19	29	800
AP 160–161	21	31	800

*Can carry 1,000 additional troops when not carrying LVT's or DUKW's.

AUXILIARY REPAIR SHIP (BRITISH LSE—LANDING SHIP, EMERGENCY REPAIR) ARL

Operational Use: Designed to provide extensive repair facilities for damaged landing craft in a landing area.

U. S.	U. S.
ARL 1, ACHELOUS.	ARL 8, EGERIA.
ARL 2, AMYCUS.	ARL 9, ENDYMION.
ARL 3, AGENOR.	BRITISH
ARL 4, ADONIS.	LSE 1 (ARL 5).
ARL 7, ATLAS.	LSE 2 (ARL 6).

US

Description Converted from LST hulls. Few alterations were necessary.

1. Deckhouse extended forward for additional workshops.
2. Athwartship 50-ton derrick, two 10-ton booms, kingposts, and winches added to deck.
3. Elevator changed into hatch.
4. Ramp removed, bow sealed.
5. Tank deck changed to include electrical, machine sheetmetal, shipfitter, blacksmith, and pipe shops, and store space.

Capacity 2 LCV (P), 10 Balsa Floats.

Armament One 3''/50, two 40 mm. quads, eight 20 mm.

All other statistics similar to LST.

LANDING SHIP, VEHICLE LSV

Converted to carry amphibious vehicles and troops.

Designation as AP being changed. AP 106, 107, are ex-CM 6, 7 (Catskill, Ozark). AP 160, 161, 108, 109 are ex-AN 1, 2, 3, 4 (Monitor, Montank, Osage, Saugus).

Length o. a., 451'–453'
Draft, (light), 21'
Beam, 60'
Speed, 19½ kts.

US

LSS LANDING SHIP, STERN-CHUTE

PRINCESS IRIS DAFFODIL

British conversions of cross-channel train ferries, designed to transport and launch loaded LCM's. The landing craft are carried on the main deck in trolleys and launched singly down a stern slipway.

Dimensions	350′ o. a. x 58′6″ x 10′11″ (mean).
Displacement	2,500 tons.
Capacity	Thirteen LCM (1)—weight less than 16 tons each—or 9 LCM (3).
Endurance	1,265 @ 11½ kts. (max.), 1,656 @ 9 kts.
Armament	Four 2 pdr., five 20 mm., five .303 Lewis guns.

UK

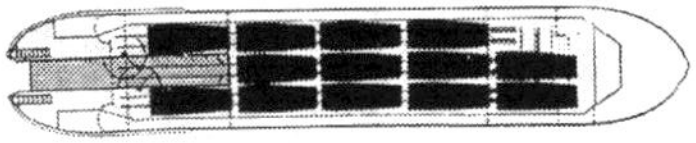

LSC LANDING SHIP CARRIER, (Derrick Hoisting)

EMPIRE CHARMAIN EMPIRE ELAINE

Operational use	To transport landing craft from the port of embarkation to the transport area.
Description	Converted British cargo carriers fitted with heavy-lift derricks.
Capacity	21 LCM; 15 officers, 280 enlisted troops.
Dimensions	Length, 416′. Beam, 67′. Draft, 26′7″ (mean).
Speed	10 Knots (cruising) Displacement 14,500 tons. Endurance 2,300 miles.

UK

▲ LSC LSG ▼ ▶

DEWDALE DERWENTDALE ENNERDALE

British oilers converted to transport and debark loaded LCM's in a landing operation. The landing craft are stowed on deck and lowered by means of overhead gantry cranes.

Dimensions	460′ (b. p.) x 59′3″ x 27′6″ (loaded).
Displace-ment	5,450 tons (light), 16,750 tons (loaded).
Capacity	15 LCM (1); gantries can carry LCM and 10 tons load.
Endurance	13,000 miles @ 12½ kts. (max.). 10,000 miles @ 11 kts. plus 5,000 miles with auxiliary.
Armament	One 4″.7, one 12 pdr., three 2 pdr., four 20-mm., Dewdale; two .303, Ennerdale; two 20-mm, Derwentdale.

Plan shows LCM stowage arrangement, photos below show gantries.

UK

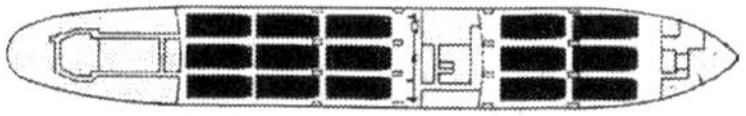

LSI(H) LANDING SHIP, INFANTRY (HAND-HOISTING)

Small British combat-loaded attack transports, classed separately because landing craft are operated from handhoisted davits.

ROYAL SCOTSMAN Class

Royal Scotsman.
Royal Ulsterman.
Ulster Monarch (bottom right).
Amsterdam.
Ben-My-Chree.
Biarritz.
Duke of Argyll.
Duke of Rothesay.
Isle of Guernsey.
Lairds Isle.
Victoria.
Duke of Wellington.
Brigadier.
Invicta (top right).
St. Helier (center).
Isle of Thanet.
Canterbury.
Princess Margaret.
Princess Maud.
Viking.
Lady of Mann.
Maid of Orleans.
Mecklenburg.

UK

For statistics on these ships see tables.

▲ROYAL ULSTERMAN ▼DUKE OF WELLINGTON

LANDING SHIP, HEADQUARTERS **LSH**

Large British ships converted to serve as combined operations headquarters. Similar U. S. ships are AGC.

Dimensions	BULOLO	LARGS	HILARY
L. oa.	412′	339′6″	420′
Beam	58′	51′4″	56′
Draft	22′ (loaded)	20′3″	23′10″

Capacity: Four LCP(L); Hilary, six LCP(L).
Armament: One 6″, one 3″ A.A., two 2 pdr. mult., eleven 20-mm.

Another unit, CITY OF EDINBURGH converting.

◄ BULOLO
▲ LARGS ▼ HILARY

UK

Photos show units of PRINCE CHARLES class.

◄ **LANDING SHIP, INFANTRY (MEDIUM) LSI (M)**

QUEEN EMMA Class
- QUEEN EMMA.
- PRINCESS BEATRIX.

PRINCE HENRY Class
- PRINCE HENRY.
- PRINCE DAVID.

Medium-sized British combat-loaded transports converted from Dutch cross-channel ferries.

Dimensions QE, 380′0″ o.a. x 47′ x 15′1″ draft.
PH, 366′4″ o.a. x 57′1″ x 18′6″ draft.

Capacity QE, 6 LCA or 6 LCS (M) and 2 LCM(1).
PH, 5 LCP (L) or 5 LCA and 1 LCS (M) and 2 LCM(1), or (3).

UK

LANDING SHIP, INFANTRY (SMALL) LSI (S)

PRINCE CHARLES Class
- PRINCE CHARLES
- PRINCE LEOPOLD
- PRINSES ASTRID
- PRINSES JOSEPHINE CHARLOTTE

PRINS ALBERT Class
- PRINS ALBERT
- PRINCE BAUDOUIN

Dimensions 360′ (PC) 370′ (PA) x 46′ x 13′1″, 13′9″ draft.

Capacity 8 LCP (L) or LCA or LCS (M).

For further information see tables.

UK

◄ PRINCE CHARLES class ▼

▲ PRINS ALBERT

LSI (L)

▲ LLANGIBBY CASTLE

▼ GLENGYLE, a unit of the Glenroy Class.

▲ KEREN ▼ MANOORA, shown prior to conversion.

LANDING SHIP, INFANTRY (LARGE) LSI (L)

Large British combat-loaded transports, equivalent to our APA's.

◄ MARNIX VAN ST. ALDEGONDE.

GLENROY Class.
- GLENROY.
- GLENEARN.
- GLENGYLE.
- MANOORA.
- KEREN.
- WESTRALIA.
- KANIMBLA.
- CIRCASSIA.
- LLANGIBBY CASTLE.
- DERBYSHIRE.
- DURHAM CASTLE.
- CLAN LAMONT.

PAMPAS.
EMPIRE BATTLEAXE.
EMPIRE HALBERD.
BARPETA.
DEVONSHIRE.
DILWARA.
DUNERA.
EL HIND.
EMPIRE ANVIL.
EMPIRE ARQUEBUS.
EMPIRE BROADSWORD.
EMPIRE CROSSBOW.

EMPIRE CUTLASS.
EMPIRE GAUNTLET.
EMPIRE JAVELIN.
EMPIRE LANCE.
EMPIRE MACE.
EMPIRE NIGER.
EMPIRE RAPIER.
EMPIRE SPEARHEAD.
EMPIRE TUGELA.
LLANSTEPHAN CASTLE.
SOBIESKI.
WINCHESTER CASTLE.

For data on these ships see the statistical table.

UK

LANDING SHIP, PERSONNEL LSP

Four miscellaneous large passenger types employed in operations similar to the U. S. AP's.

Batory, shown above is a typical example of this group. Many other large liners have also been equipped with landing craft, but have no official designation.

◄ CIRCASSIA

UK

AGC AMPHIBIOUS FORCE, FLAGSHIP

U. S. Navy equivalent to the British LSH. These ships serve as control headquarters for landing operations; are readily distinguished by their heavy radar mast. Two Maritime Commission designs make up this type—the ANCON (below), a C3-P & C, and the APPALACHIAN Class, a C2-SB1 design (right).

APPALACHIAN Class

Appalachian.	AGC 1.	Blue Ridge.	AGC 2.
Rocky Mount.	AGC 3.	Catoctin.	AGC 5.
ANCON	AGC 4.	Building	AGC 7-14.

	AGC 1–4.	AGC 5.
Dimensions	459'3"x63'	493'x64'
Displacement (loaded)	11,998 tons	13,144 tons
Boats	4 LCV (P) 1-2 LCP(L)	8 LCV(P)
Speed	15.5 kts	18 kts.
Armament	2-5"/38 2-40 mm. 18-20 mm.	2-5"/38 4-40 mm. 10-20 mm.

Accommodations 130 officers, 869 enlisted

US

HIGH SPEED TRANSPORTS (DE CONVERSION) APD

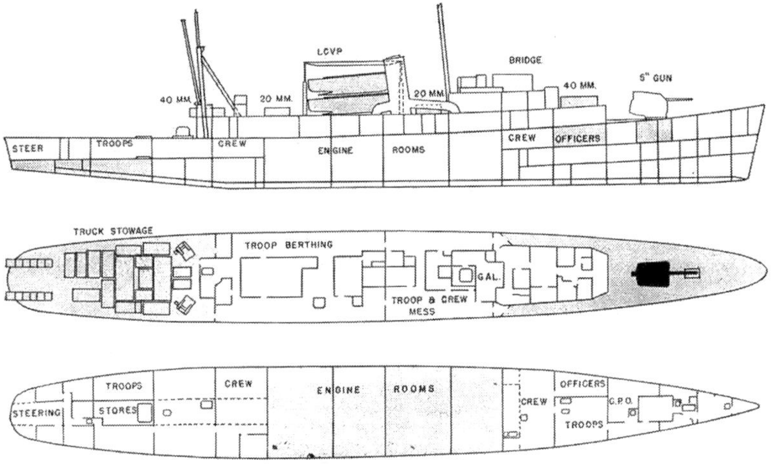

Operational use	To transport and land raiding or reinforcing troops secretly.	
Description	Plans are being developed for the conversion of DE hulls to serve as APD. The WGT (Westinghouse Geared Turbine) type is being used. No class numbers have been assigned yet.	
Capacity	4 LCVP at davits 162 Troops 6–1/4 ton trucks 2–1 ton trucks 4 carts T4EI	4-75 mm. Pack Howitzers 4,500 cubic feet cargo ammunition 1,000 cubic feet gasoline 3,500 cubic feet general stores
Dimensions	Length 306′, Beam 37′	
Draft	14′9″ (full load), 16′1″ (with sound dome down)	
Displacement	2,043 tons (full load)	
Armament	1–5″/38 cal. 3–40 mm. twins 8–20 mm.	
Speed	23 kts. (maximum)	
Endurance	5,000 mi. @ 15 kts. 2,000 mi. @ 23 kts.	
Propulsion	Geared turbines	
Crew	212	

US

Photos show LCP(R) on davits. No larger craft can be carried on these ships.

Note camouflage on Manley (above), designed for inshore work.

HIGH-SPEED TRANSPORTS (DESTROYERS) **APD**

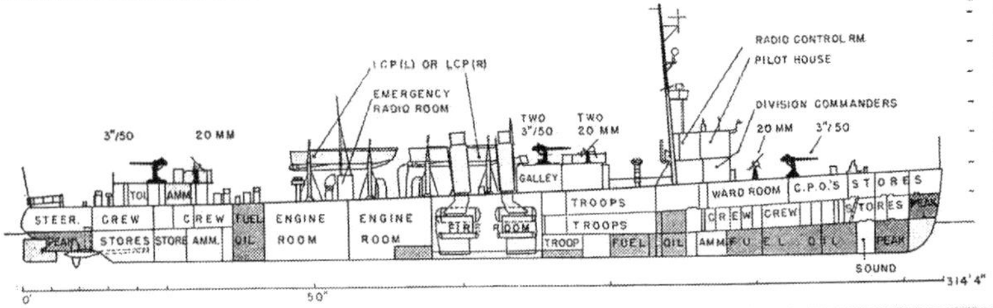

Converted "flush-deck" destroyers designed to accommodate raiding parties and land them by small craft. Nineteen are now in service, others are contemplated for conversion.

Dimensions	314'4" o.a. x 31'8"; 11'11" plus 4' sound dome (max. draft).
Displacement	1,060–1,090 tons; 1,700 (loaded).
Capacity	Four LCP(L) or LCP(R) and one Marine rifle company. Over 200 men have been successfully accommodated. 140–148 berths.
Armament	Four 3" .50-cal. DP; five 20-mm., 6 D. C. projectors are carried.
Speed	23 kts. (max.). Crew, 112.

US

APc COASTAL TRANSPORT (SMALL)

Small wooden transports designed for use around advance bases, in raiding parties, or light-draft conditions in conjunction with large landing operations. Many of these craft were previously used as AMc, coastal minesweepers. The type illustrated is a new class, designed for the purpose.

Capacity	Troops, 2 officers, 60 enlisted; or 17 tons of cargo in forward hold.
Dimensions	103′ (wl) x 21′3″ x 11′ draft.
Displacement	238 tons.
Endurance	3,095 miles @ 9.5 knots (cruising).
Propulsion	One 400-h.p. direct-drive Diesel.
Armament	Four 20-mm.
Crew	3 officers, 17 enlisted.

US

ISBN #978-1-940453-16-3

84414954R00075

Made in the USA
Middletown, DE
19 August 2018